DR RAMAN PRINJA

WONDERS OF THE PLANETS

Visions of our solar system in the 21st Century

MITCHELL BEAZLEY

To my three celestial lights: Kamini, Vikas, and Sachin

Wonders of the Planets

by Dr Raman Prinja

First published in Great Britain in 2006 by Mitchell Beazley,
an imprint of Octopus Publishing Group Limited.

Commissioning Editor **Jonathan Asbury**
Senior Editor **Peter Taylor**
Design **Colin Goody**
Production **Gary Hayes**
Picture Research **Guilia Hetherington**
Copy Editor **Jemima Donne**
Proofreader **Suzanne Arnold**
Indexer **Sue Farr**

ISBN-13: 9 78184533 244 0
ISBN-10: 1 84533 244 X

A CIP catalogue record for this book is available
from the British Library

Typeset in Gill Sans and Ticket

Reproduction by Chromographics

Printed and bound in China

Previous page
Saturn's icy moon Dione contrasts against the colourful atmosphere of the
giant gas planet in this striking image relayed from the Cassini spacecraft.
A rich range of planetary features can be seen, including craters and cracks
on the terrain of Dione and oval-shaped storms in Saturn's upper layers.

Extensive radar observations taken over four years here provide a more
detailed global map of the surface of the totally cloud-enshrouded planet
Venus than we currently have of the Earth's ocean floors. This false-colour
radar mosaic of Venus is based mainly on Magellan spacecraft observations,
and has an impressive effective resolution of 3km (1.9 miles). The purple to
green colour codes represent higher and lower elevations, respectively.

1990-1994

Contents

Introduction

We are today in the midst of a period of truly unprecedented discovery and exploration of the solar system. Advanced spacecraft missions have combined with robotic landers and powerful telescopes to reveal remarkable details of an incredibly diverse and beautiful system of planets, moons, and rings. Through increasingly sophisticated feats of space engineering, vehicles have also been expertly manoeuvred to land on asteroids, gather material from the sweeping tails of comets, and sample the wind particles blowing from the Sun. These advances have been complemented by an improved understanding of our own planet, including a greater appreciation of the wide and severe conditions under which life can exist on Earth. This realization has led to the flourishing new subject area of astrobiology and the push to explore other locations in the solar system where conditions may be conducive for microbial life to thrive. On a grander scale, the rapidly growing discoveries of planets orbiting other stars have, for the first time, placed us in a position where we can ask profound questions about the origin and uniqueness of our solar system.

One of the most important disciplines of modern solar system astronomy is the comparison of different planets with the Earth, and with each other. Known as comparative planetology, the collective study of planets and other large bodies in the solar system can teach us a great deal about the history and future of the Earth.

Exploration of the solar system is dominated by spacecraft missions (this includes orbiters, probes placed onto planetary surfaces, and relatively brief flybys). From top to bottom, these are images from several spacecraft showing the terrestrial, or rocky, planets Mercury, Venus, Earth (with the Moon), and Mars – shown to scale with each other. The lower four are the giant gas planets Jupiter, Saturn, Uranus, and Neptune; these are not to scale.

Comparing the similarities and differences in the geology and atmospheres of the planets helps us to understand the alien worlds and the place of our own world in the solar system. Rocky bodies such as Mars, Mercury, and the Moon have, for example, preserved records of their histories in craters and ancient lava flows. In contrast, Venus, with its heat-trapping atmosphere of carbon dioxide, gives us an insight into a possibly hostile future for the Earth.

This comparative approach is adopted throughout this book as we explore a solar system that is rich in wonders, from major geological phenomena, such as impact craters and volcanoes, to stormy atmospheres and glorious ring systems. In addition, now that more environmental conditions are regarded as habitable on Earth, we are exploring the potential habitability of other planets and moons. Ultimately, our solar system as a whole can be compared with the multiple planets of other star systems.

An inventory of the solar system

The bodies in the solar system come in a variety of sizes, colours, and environments. They range from vast globes of gas to small balls of rock, and from frigid ice-encrusted moons to piles of rubble loosely held together by a weak "self-gravity". The solar system comprises our very average star, the Sun, and all the bodies that orbit around it. The major constituents of the solar system are the nine planets, the numerous satellites or moons of the planets, and the swarms of small objects such as comets and asteroids.

If viewed from well above the Sun's north pole, all the planets orbit the Sun in the same counter-clockwise direction. With the exceptions of Mercury and Pluto, the paths of the planets are nearly circular and confined to a flat plane that is only marginally tilted from the plane marked by the Sun's equator. These orderly properties are important because they provide direct constraints on how the planets were formed billions of years ago. The solar system is not only incredibly large, but also extremely empty. The planets are very small in comparison to the distances between them. As a model, imagine that the scale of the solar system is reduced by a factor of 15 billion and the Sun is represented by a light bulb about 10cm (4in) across. The largest planet Jupiter would then be represented by a 1cm (0.4in) marble, placed 52m (170ft) away. On this greatly reduced scale the Earth would be a 1mm speck about 11m (35ft) from the light bulb. At its farthest, Pluto, the most distant planet, would be less than the span of a full-stop on this page, at a distance of 317m (1040ft) from the "Sun". If you were to include the next nearest star system, Alpha Centauri, in this scale model you would need to place another light bulb almost 2900km (1800 miles) away.

On a larger scale, it is important to note that our Sun is just one of 200 billion stars held together by gravity to form our spiral-shaped Milky Way Galaxy. The Sun is a typical component of the Galaxy, located in an insignificant position about 26,000 light years from the centre. (One light year is the distance travelled by light in one year, which equals about 9460 billion km or 5880 billion miles.) At this location, it takes the Sun

Our backyard star, the Sun, dominates the solar system with its great mass and luminous power. The dynamic and violent aspect of the Sun is beautifully captured in this image from the Solar Heliospheric Observatory (SOHO) in space. A vast eruptive prominence (top right) is seen speeding away at 75,000km/h (45,000mph), as it forms a gargantuan structure more than 50 times the Earth's diameter.

12-2005

and its family of planets about 225 million years to orbit just once around the Galaxy. Furthermore, our Milky Way Galaxy has no special location among the gathering of about 25 galaxies that forms a gravitationally bound group in our "corner" of the Universe. These neighbouring galaxies form just a minuscule fraction of the 100 billion galaxies of various configurations in the observable Universe. Each of these galaxies is in turn loaded with millions or billions of stars, in an unimaginably vast and diverse cosmic scene.

A brief tour of the solar system

The Sun dominates the solar system. It contains more than 99 per cent of the total mass and thus provides the major gravitational field to hold the planets in their orbits. The Sun's radiation provides the primary source of heat and the light reflected

by the planets. The planets do not generate any light of their own.

Moving away from the Sun, the first planet, Mercury, is difficult to observe from Earth since it is always low on the horizon. The Mariner 10 spacecraft visited Mercury in 1974 and returned images of a heavily cratered visage very reminiscent of Earth's Moon. The crusty surface of this relatively small, rocky planet is also marked by vast cracks and faults that were created when the body was still cooling and compressing after its formation. Mercury has no atmosphere or moons and is the densest of the planets due to its relatively large iron core.

Venus is also difficult to study from Earth because its surface is totally shrouded by its thick, mainly carbon dioxide, atmosphere. Though the planet is similar to the Earth in size and volume, its dense atmosphere creates a crushing surface

Our Sun, with its family of planets, is just one of the 200 billion stars that make up our Milky Way Galaxy. A fraction of this flattened, spiral-shaped system of stars and gas clouds is seen edge-on in this panoramic view obtained by NASA's Spitzer Space Telescope. Spitzer uses infrared cameras to "see" through obscuring matter in interstellar space to reveal a rich scene that includes star- and planet-forming clouds, youthful stars, and more evolved stars nearing the ends of their lives.

pressure that is more than 90 times that on Earth. Venus' hostile atmosphere also creates an extreme greenhouse effect that traps radiation from the Sun and heats the surface to almost 480°C (896°F). Spacecraft observations have revealed vast plains of ancient lava, and evidence that Venus endured catastrophic volcanic activity several hundred million years ago.

Our home planet is most remarkable for its unique oceans of surface liquid water and the fact that it is teeming with such an incredible variety of living species. The Earth is currently the most geologically active planet in the solar system, with its internal heat powering volcanoes, and movement of its surface crust creating destructive earthquakes. The Earth is also unusual in that it is paired with a moon that is relatively large compared with the size of its parent planet.

The red planet Mars, named after the Roman God of war, is the most thoroughly studied planet beyond Earth. A series of orbiting spacecraft, probes, and surface landers has been sent to the planet over the past three decades. Telescopic observations on Earth have also enabled long-term patterns to be analysed, including the seasonal polar ice caps and planet-engulfing dust storms. Mars has a thin carbon dioxide atmosphere and an extremely cold average surface temperature of -63°C (-81°F). This fascinating world is marked by enormous extinct volcanoes, deep canyons, and evidence of a wet past. It also remains possible that Mars harboured primitive bacterial life-forms in the past, which may even still be present today.

The next planet, Jupiter, is the largest in the solar system and it is the innermost of the four giant gas planets. Jupiter has a

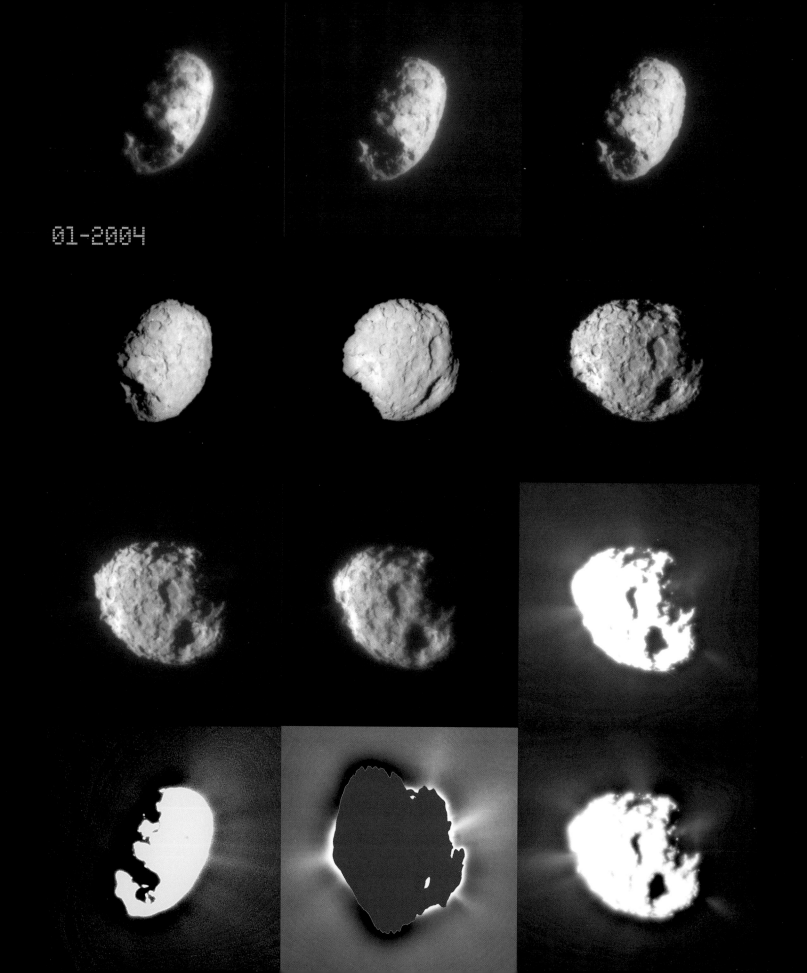

swirling upper atmosphere that bears the scars of a very dynamic weather system, which includes fierce hurricanes and zonal jet winds. It has a substantial system of moons, several of which are fascinating worlds in their own right. The latest exploration of this king of planets was conducted during the late 1990s by the Galileo orbiting spacecraft.

Saturn is the second-largest gas giant, with an overall structure that is broadly similar to Jupiter's. Though all four gas planets are encircled by rings, none other has the brilliance and shear beauty of the rings around Saturn. A few of the planet's numerous moons nestle within the rings and are now known to play a direct role in carving out complex structures in the material around them. Superb new views of Saturn are currently being delivered by the Cassini spacecraft mission, which is the first to go into orbit around the planet.

Farther out in the solar system, Uranus takes 84 Earth years to complete a single orbit of the Sun. It was discovered by the English astronomer William Herschel in 1781, but our closest views of Uranus were obtained only in the mid-1980s when the Voyager 2 spacecraft flew past it. Uranus is peculiar for the tilt of its axis of rotation compared with the other planets'; it seems to be lying over on its side. This gas planet is orbited by a system of more than 20 moons and a delicate set of rings.

Uranus is twinned with the outermost gas giant, Neptune. However, the latter has a more active weather system, with powerful winds and cyclone-like storms. Neptune's family of at least 10 moons and intricate rings was visited by the Voyager 2 spacecraft in 1989.

Pluto is the planet farthest from the Sun and takes 248 years to complete one lap of its orbit. It is a small body made of ice and rock, very reminiscent of some of the satellites of the giant gas planets. This similarity, and the recent discovery of other comparable objects orbiting beyond Neptune, has precipitated renewed discussion that questions whether Pluto should in fact be classified as a planet.

The origins of the solar system

The formation of our solar system took place about four and half billion years ago, with the collapse under its own gravity of a vast cloud of cold gas, dust, and ices. The combined actions of gravity and a slow spin gradually transformed this ball-like nebula into a plate-like disc of thinly spread material. The central region of the disc became very highly compressed, with the gas being held in conditions of incredibly high pressure and temperature. Eventually this hot core yielded the birth of the Sun, which, like all stars, was (and still is) powered by the process of nuclear fusion.

Meanwhile, over millions of years, the tiny solid particles left-over in the surrounding disk bumped into one another and stuck together, growing ever larger as their gravitational forces increased. Kilometre-sized bodies called planetesimals formed. These then grew further through collisions to become the nuclei of planets or of smaller bodies such as asteroids and comets. Near the youthful Sun, where the temperature remained very high, denser metals and rocks solidified to form the terrestrial planets, Mercury, Venus, Earth, and Mars. Because

NASA's Stardust spacecraft has recently completed a mission to collect dust particles from a comet called P/Wild 2 and return the samples to Earth. The images shown were taken during the closest approach to the 5.4km (3.3 mile) wide comet. The spacecraft swooped to within 240km (149 miles) of the surface revealing a pock-marked terrain. The particles are unchanged from when they formed 4.5 billion years ago. Analysis will teach us about the formation of the solar system.

05-2005

of the rarity of raw materials in these inner regions, there was not enough matter to assemble planets much larger than the Earth. The outer regions of the solar system were cold enough for abundant ices to condense. Mixed with smaller portions of other solids, these "blocks" of icy matter were able to grow into substantial balls, perhaps 10 to 20 times the size of the Earth. The greater gravity of these trigger cores subsequently attracted large concentrations of hydrogen and helium gas, which made them grow even bigger. At the completion of this process, the four giant gas planets Jupiter, Saturn, Uranus, and Neptune had been created in the outer solar system. These vast bodies were also able to attract numerous moons, and subsequently all four gas giants have also formed ring systems. Finally, the planet status of Pluto continues to be debated because this peculiar body is most unlike the other eight major members orbiting the sun.

The new golden age of exploration

The discoveries and spectacular imagery presented in this book derive from some of the most ambitious spacecraft missions and advanced telescope facilities available in recent years. Between 1995 and 2003, for example, the Galileo spacecraft orbited Jupiter, monitoring not only its stormy atmosphere, but also its highly magnetic environment and geologically diverse moons. Since 2004, the Cassini mission has been studying Saturn in magnificent detail, beaming back astounding images of the planet, its exquisite rings, and its wondrous satellites. This mission also launched the Huygens probe that successfully landed on the giant moon Titan in January 2005.

Mars has been inundated by visitors from Earth. Since the mid-1990s, NASA's Mars Global Surveyor and the European Space Agency (ESA)'s Mars Express spacecraft have been exploring the surface features, atmosphere, and polar ice caps of the dusty planet. In 2004 NASA's two robust Mars rovers started roaming the planet, trekking up to 100m (110yd) across the surface every martian day.

Parked on the surface of Mars, the Exploration Rover Spirit beamed back this stunning scene of the Sun setting below the rim of a crater. Large amounts of high-altitude dust help scatter sunlight to the night side of the planet. The twin Mars rovers have spent more than a full martian year (687 Earth days) gathering evidence about the ancient environment.

Over the past few years two missions called the Solar Heliospheric Observatory (SOHO) and Ulysses have provided unique observations of the Sun. They have monitored our star's dynamic atmosphere, tracking huge eruptions of electrified particles that travel over a great region of interplanetary space. More ingenious spacecraft have also been launched, including Stardust, which collected dust from the tail of Comet P/Wild 2 in January 2004 and returned it to Earth in January 2006.

Our knowledge of the solar system is also being advanced by telescope observations that exploit powerful facilities such as the 8–10m-(26–33ft) aperture telescopes perched on the Andean peaks in Chile and the volcanic mountains in Hawaii. NASA's "Great Observatories" in space are providing superb quality images and enormous volumes of valuable data. They include the Hubble Space Telescope, Spitzer that "sees" in the infrared waveband, and the Chandra observatory that captures high-energy phenomena in sharp X-ray pictures.

This books aims to mark the current period of pioneering exploration. The new "golden age" of exploration is one of breathtaking discoveries that are challenging scientific thinking and raising profound questions. The latest findings not only impact on the context of our planet in the solar system, but also the place of our solar system in the Galaxy and the Universe.

Our knowledge and understanding of the solar system is being advanced today by remarkably detailed observations of the planets, moons, and other smaller bodies obtained using powerful telescopes on Earth and sophisticated robotic spacecraft. Shown here is the 8m (26ft) class Gemini North telescope on the peaks of Mauna Kea, Hawaii.

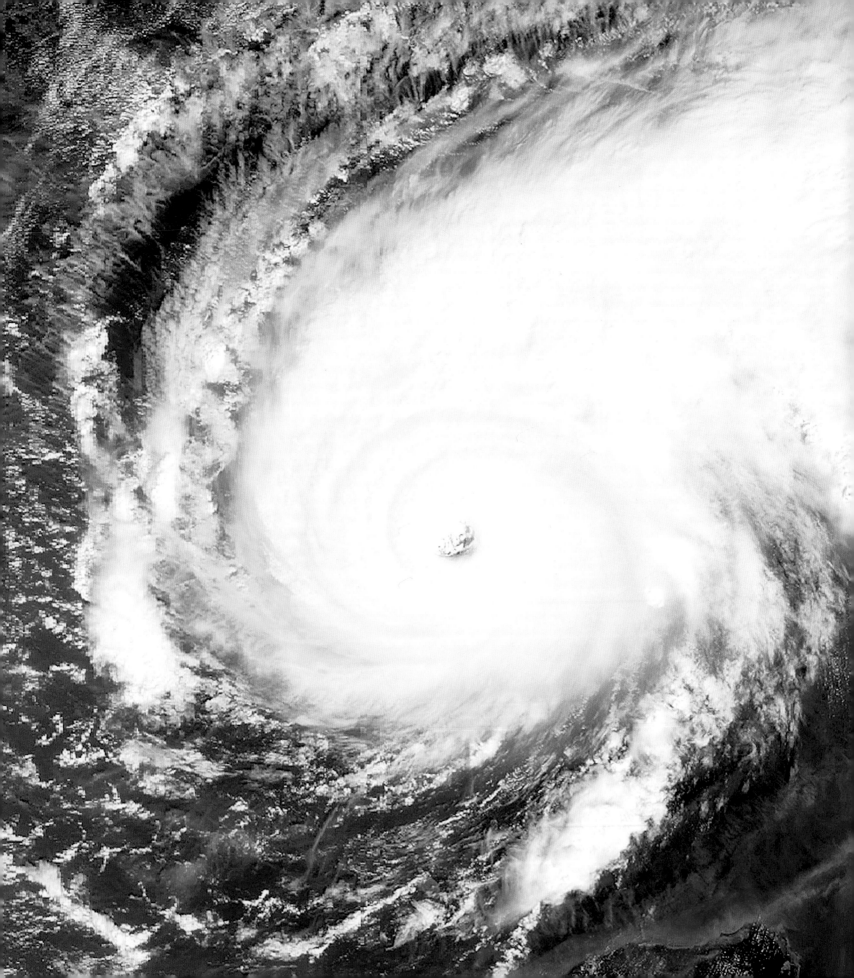

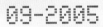
09-2005

The power of Hurricane Rita as it approaches the southern Florida coast is captured in this photograph taken by the European Space Agency's Envisat satellite. The details evident here include rough ocean waves and intricate structures within the ominous swirling clouds. Satellites such as Envisat are equipped with advanced radar-imaging instruments that provide vital information on the wind strength and evolution of hurricanes.

Hurricanes and Storms

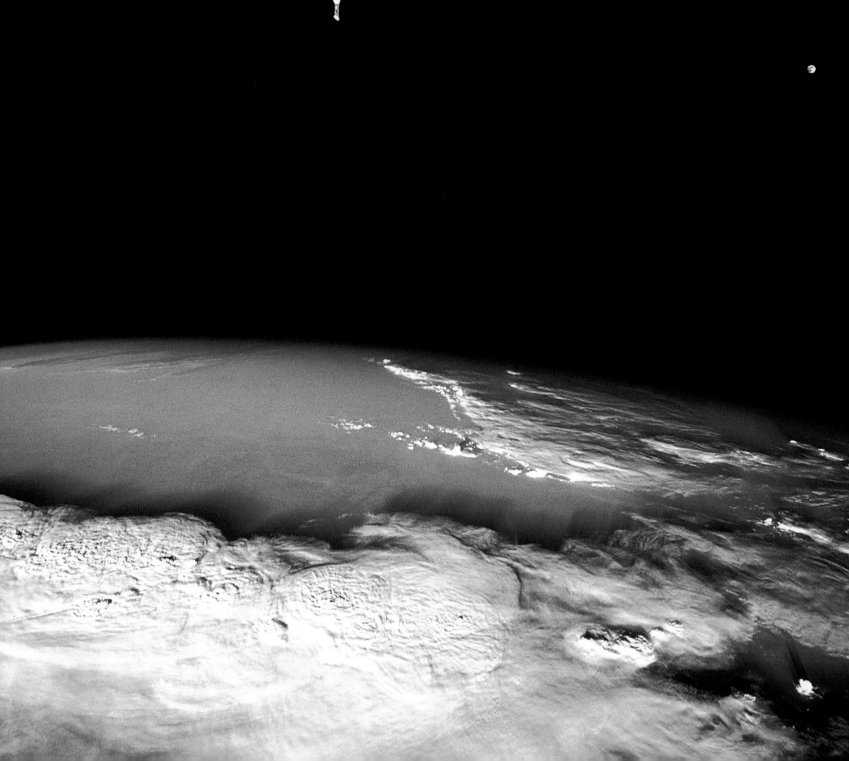

You might think that the weather where you live is wild and unpredictable, but rest assured that the extreme conditions on other planets in the solar system are beyond anything we experience on Earth. We will see in this chapter that even the angriest storms on Earth pale in comparison with the cyclones that rage on giant gas planets such as Jupiter and Saturn. These extra-terrestrial storms can persist for months, years, and even centuries. Desert winds on Earth are contrasted here to phenomenal dust storms that can engulf entire hemispheres of planet Mars, and we witness super-thunderstorms on Jupiter that can deliver lightning bolts that dwarf those we see on Earth. The weather extremes are not limited to planetary bodies in the solar system, but extend to huge eruptions of electrified gas from the Sun that travel through space to produce spectacular glowing aurorae in the atmospheres of planets. However, observing weather in the solar system is more than just a thrill. Weather follows the movement of matter and energy. Scientists study weather on other planets to learn more about their atmospheres and to predict conditions on the surfaces of planets such as Mars; these predictions are a prerequisite for human exploration of the red planet. In addition, the study of long-lived weather patterns on other planets can also help us to understand Earth's own erratic weather.

Astronauts aboard the Space Shuttle Discovery shot this fantastic picture of the Earth and its turbulent atmosphere. The view below is of gathering heavy thunderstorms. Floating above the planet is Russia's Mir space station. The Earth's natural satellite, the Moon, is also seen, on the right. Data obtained on individual hurricanes from Space Shuttle flights are combined with measurements from other sources, such as aircraft and satellites, to improve our ability to predict the behaviour of storms.

A Brief Tour of the Planets' Weather

Climate and weather are very familiar concerns for us on Earth as we plan our holidays and weekend trips. In a particular place and time, weather conditions may include rain, wind speeds, cloudiness, areas of high pressure, or cold fronts. Very occasionally we may also have to take account of more energetic phenomena such as tornadoes and thunderstorms. On a wider scale, the regional climate is the typical weather of a country or place on Earth. We can also think of global climate as the climate of the Earth as a whole, accounting for variations between different regions. Any planet in the solar system that has a substantial atmosphere can have weather. Indeed, as our concerns grow over how humanity might induce changes in Earth's atmosphere, and thus alter global weather patterns, scientists can learn a great deal about the most critical factors by studying the weather on other planets.

The terrestrial, or rocky, planets Earth, Mars, and Venus have different atmospheres and therefore different weather. In greater contrast, the giant gas planets Jupiter, Saturn, Uranus, and Neptune are surrounded by vast gaseous envelopes that generate weather on a much grander scale. Generally, the weather on planets may be affected by the tilt of a planet's axis (which causes seasons), its distance from the Sun, the length of its day, and the nature of its atmosphere. For example, on Earth heat radiation from the Sun is the primary source that combines the atmosphere and key elements such as pressure, wind, and moisture, to make weather. On the gas planets Jupiter and Saturn, a second driving force for weather is the heat generated by energy released from within the planets themselves.

Before taking a look at some of the most awesome planetary weather systems, let's look some of the more general conditions on the planets in our solar system. For example, because the planet Mercury has essentially no atmosphere, its weather changes are not seen as storms, but rather as the most extreme swings in temperature that range between -173°C (-280°F) at night and 425°C (800°F) during the day. Although Venus has a similar mass and size to the Earth's, it's entirely shrouded by a thick atmosphere of carbon dioxide and sulphuric acid clouds. The dense atmosphere traps any energy from the Sun that reaches the planet's surface, thus producing the runaway greenhouse effect. This has made the surface of Venus one of the hottest places in the solar system, with temperatures reaching around 460°C (865°F). The planet Mars, on the other hand, has a very thin carbon dioxide atmosphere, barely one per cent as dense as Earth's at sea level. Mars' orbit around the Sun, however, is less circular than Earth's, which means that when Mars is closest to the Sun, parts of the planet can enjoy 40 percent more sunlight. This variation can lead to dramatic seasonal changes on Mars, such as dust storms and shrinking polar caps.

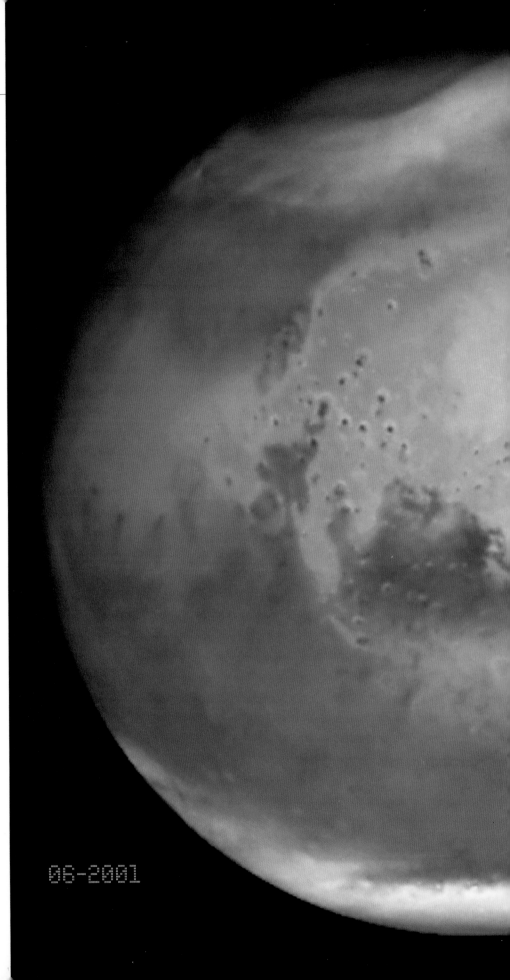

NASA's Hubble Space Telescope has been a powerful instrument for monitoring long-term weather patterns on Mars. In this image, the rise of a new dust storm is caught near the planet's icy northern cap. Less than three months later, a phenomenal amount of airborne dust had virtually obscured all the Martian surface features seen here.

Moving farther out, the axis on which Jupiter spins is tilted by only three degrees (compare this with Earth's tilt of 23.5 degrees), which means there is essentially no difference between the seasons on this planet. However, Jupiter is the fastest-spinning planet in the solar system, and it has a very turbulent atmosphere. The different chemicals in its atmosphere can also lead to extreme variations in the temperature. Similar conditions are found on Saturn, although its atmosphere is less colourful than Jupiter's. Another rapid rotator, Saturn has substantial overall winds that flow from east to west, giving the planet its banded appearance. These large-scale winds are called "zonal flows" and are often accompanied by oval storm systems.

Uranus is peculiar because its axis of rotation is tilted by 90 degrees; you might say that it is tipped over on its side relative to the other planets. This orientation causes some strange seasonal effects. For almost a quarter of its orbit around the Sun (about 84 Earth years), the Sun shines over each of Uranus' poles, leaving the other hemisphere of the planet immersed in a long, dark winter.

Next up, Neptune's cloud-like features are more easily seen than Uranus' as it has a more dynamic weather pattern, together with violent winds and Earth-sized storm systems. Finally, little is known about Pluto's seasons. The tiny planet has a highly eccentric orbit around the Sun, which makes it possible that its tenuous methane atmosphere freezes and falls as ice when it is farthest from the Sun.

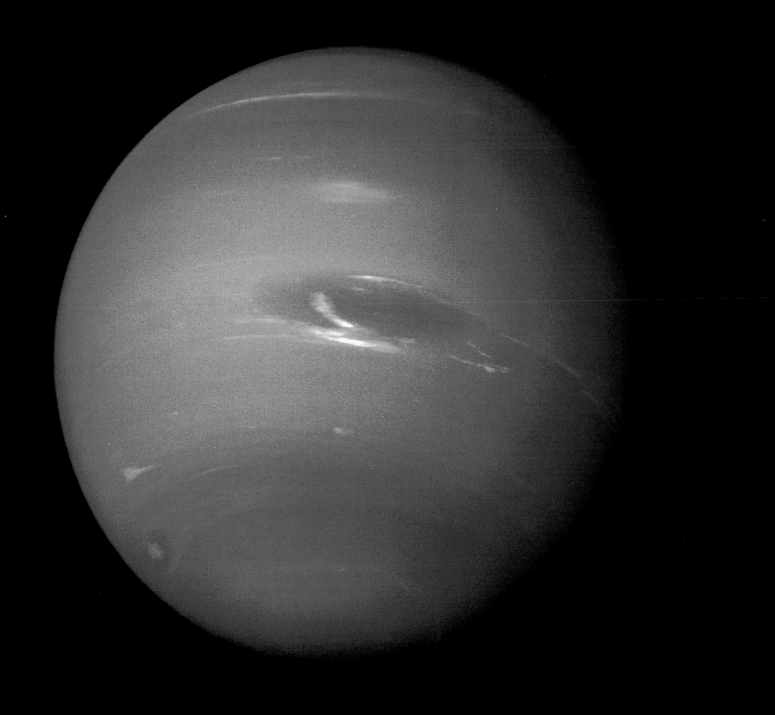

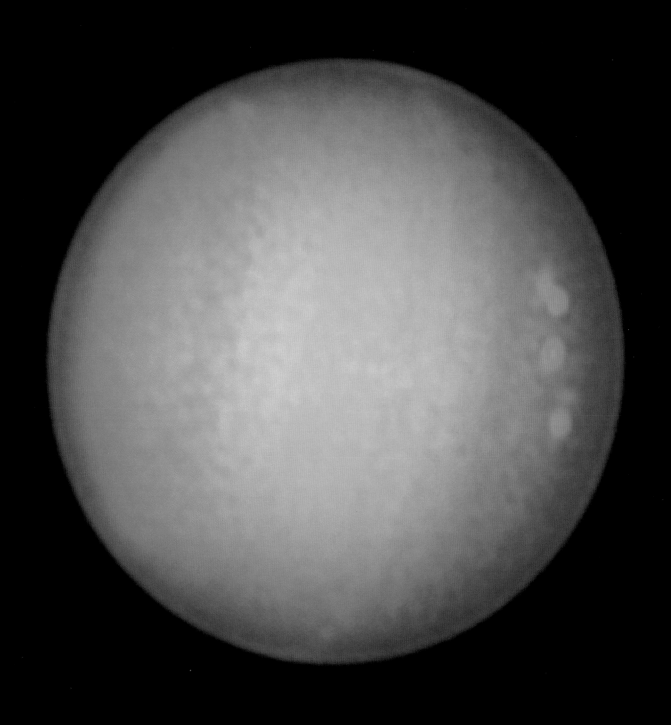
08-2003

Wind and Dust

On Earth we recognize hurricanes as one of nature's most destructive weather phenomena, often evoking a feeling of awe and mystery. Terrestrial hurricanes are born over water and driven by heat energy from the Sun that is stored in tropical oceans. The potent mixture of moisture, winds, warm oceans, and unsettled weather can combine to produce gigantic waves, high-speed winds, and torrential rains. Hurricanes on Earth are sometimes called typhoons or tropical cyclones. Whatever the name, they most often bring death and destruction. In late August 2005, for example, a very powerful hurricane named Katrina struck the Gulf coast of Mexico with winds of up to 243km/h (151mph). As it slammed into Louisiana, USA, and came inland, Hurricane Katrina brought with it an enormous surge of water that devastated communities and submerged almost 80 percent of the city of New Orleans. A comparable gigantic cyclone swept across Bangladesh in April 1991, creating a tidal surge 7m (23ft) high and resulting in the deaths of almost 100,000 people.

How hurricanes start on Earth

Tropical cyclones and hurricanes need three main ingredients at birth: a cluster of thunderstorms, warm ocean temperatures, and stable, light upper winds. These major weather systems are powered by heat. They suck in warm humid air from the lower atmosphere, which then rises and condenses. The rising air creates a region of low pressure near the ocean that draws in more warm air, feeding a continuing storm. Bands of thunderstorms form and the hurricane cloud tops rise higher into the atmosphere. Winds blowing in opposite directions immediately get the towering hurricane spinning (counter-clockwise in the Northern hemisphere and clockwise in the Southern hemisphere). The now ferocious winds combine with low pressures at the eye, or centre, of the storm to create gigantic rises in the sea level, known as storm surges. These surges, topped with powerful waves, can cause severe flooding along coastlines. Hurricanes on Earth can last almost two weeks while over the ocean. They eventually decay and lose power for several reasons. Winds can act to rip (shear) them apart, as can happen if the hurricane moves over cooler waters. Most commonly, reaching land means that a hurricane's main source of moisture is shut off and it consequently slows, weakens, and dies out.

Hurricanes are categorized according to the strength of their winds. A category-one storm reaches wind speeds of up to 153km/h (95mph), with storm surges of about 1.5m (4.9 feet) above the norm. In contrast, the strongest category-five hurricanes have wind speeds greater than 249km/h (155mph) and storm surges that raise water levels more than 5.5m (18ft). By

This image from NASA's Terra satellite shows the detailed make-up of the category-five Hurricane Isabel. With sustained winds in excess of 260km/h (160mph), this was an unusually powerful storm that maintained its category-five intensity for more than 30 hours. The hurricane eventually struck the coast of North Carolina (USA), resulting in devastation amounting to millions of dollars in property damage.

09-2003

international agreement, each year's hurricanes are given alphabetical names that alternate between male and female, with the list of names being revised every six years.

It's much worse on the giant planets

While hurricanes and cyclones have an awesome destructive power on Earth, the weather conditions on other planets in the solar system can be considerably wilder and more violent. Recent spacecraft missions such as Galileo and Cassini have provided fascinating new insights into the dynamic, highly turbulent atmospheres of the giant gas planets Jupiter and Saturn. Remember these are vast, rapidly spinning planets mostly composed of hydrogen and helium gas. Of some importance to their energetic weather systems is that both Jupiter and Saturn are overheated. That is, both these planets radiate more energy than they receive from the Sun. Jupiter generates its additional energy from a slow contraction of the entire planet that first started when it formed about 4.5 billion years ago. This heat-

generating contraction has decreased a lot today, down to about 1mm per year, in fact. The main source of Saturn's additional heat is not the contraction of the planet, but rather a result of droplets of helium raining within its deep interior. The energy of this motion is being converted to heat.

Even looking through a modest telescope on Earth, it is clear that the upper (visible) layers of Jupiter and Saturn are in turmoil. The planets reveal a banded appearance, with several oval-shaped features, and many different colours that represent the chemical make-up of the observable atmosphere. The largest and most spectacular storm in the solar system is known as the Great Red Spot and is on Jupiter. When 17th-century astronomers first turned their simple telescopes to Jupiter, they immediately noticed a prominent, reddish, oval-shaped region on the gargantuan planet. More than 300 years later, this awesome storm is still raging. At its widest the Great Red Spot is more than 24,700km (15,400 miles) across, which is almost twice the size of the Earth. Within this churning Jovian storm the

Turbulent and complex patterns are clearly evident in the clouds of Jupiter in this colour image taken by the Cassini spacecraft. The upper atmosphere of the giant planet has a more banded appearance at mid latitudes, in contrast to cyclone-like patterns near its polar region. The planet's rapid spin rate, and the heat generated within its interior are thought to power the energetic phenomena seen in Jupiter's active atmosphere.

06-1996

This false-colour image of Jupiter's incredible Great Red Spot was taken by the Galileo spacecraft. Several infra-red filters were used to penetrate the planet's upper cloud layers, revealing details of this centuries-old storm system. The blue and black regions represent deep clouds, while the pink and white areas are hazes and clouds formed higher up. The vast storm is surrounded by thunderstorms that likely feed energy to it.

wind speeds are estimated to be about 435km/h (270mph). Astronomers speculate that the Red Spot may merely be the top of a gigantic tower of rising gas that extends deep into the interior of the planet. Unlike the low-pressure hurricanes that strike the Gulf of Mexico on Earth, Jupiter's Great Red Spot is an anti-cyclone, which means it is a high-pressure storm that spins once every six days in a counter-clockwise direction. Repeated observations made by the Hubble Space Telescope over a decade have shown that the Great Red Spot changes its shape and colour, though it always remains fixed in latitude just south of the planet's equator.

So what's the driving engine that keeps the Great Red Spot going for so long? We noted earlier that the planet Jupiter generates deep-seated heat energy, which partially maintains the storm from below. Also, the Great Red Spot contains a huge amount of mass, which tends to make it more stable, and, unlike on Earth, there are no land masses and continents on Jupiter to dampen the storm. Another remarkable aspect of the Great

Red Spot noted by the Galileo satellite is the occurrence of additional storm systems very nearby that resemble clusters of thunderstorms. However, these are still considerably more powerful than any thunderstorms we know on Earth. It seems that heat leaking upwards from the planet's highly compressed centre provides the source of turbulence in Jupiter's atmosphere, forming towering super-thunderstorms that can deliver awesome bolts of lightning. These dramatic disturbances appear eventually to pour their energy into larger storms such as the Great Red Spot.

Cyclone-like ovals and spots also appear temporarily in the cloud bands of other giant gas planets, especially Saturn and Neptune. Despite the fact that Jupiter and Saturn have similar chemical compositions (including hydrogen, helium, methane, and ammonia), their appearance differs starkly. Jupiter exhibits a much more colourful atmosphere, while Saturn has less atmospheric structure. The reasons for this difference are related to the depths of their atmospheres. Jupiter's greater mass

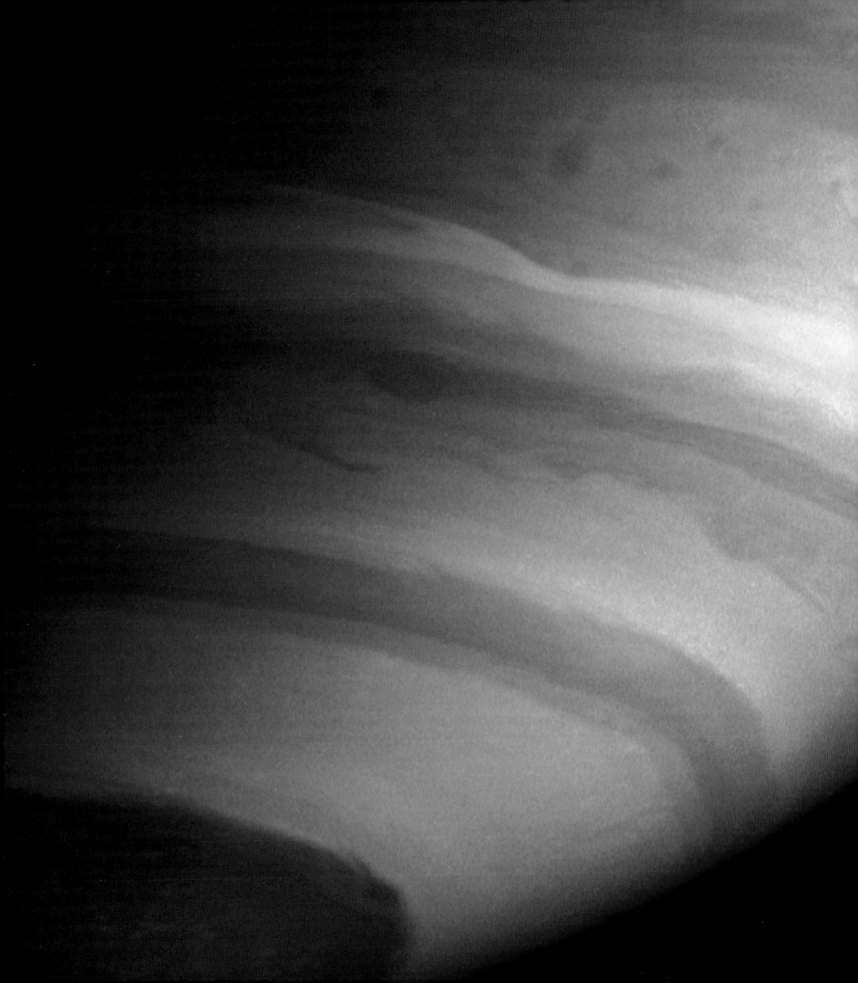

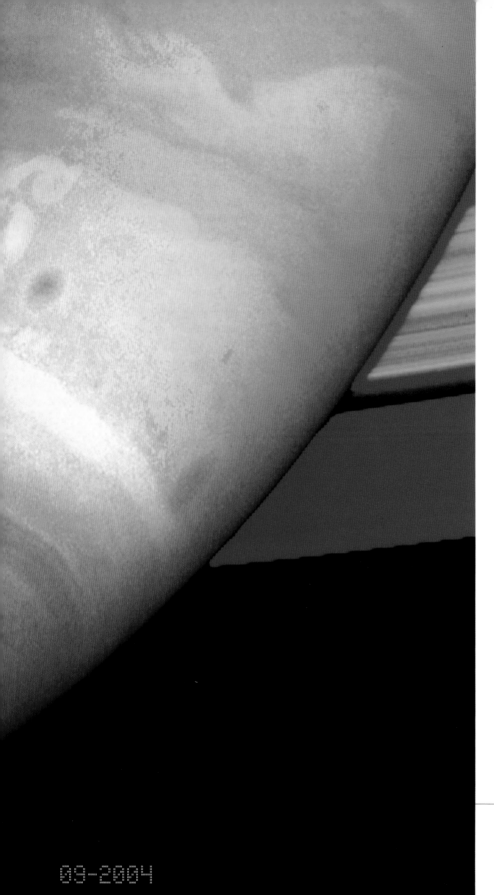

`09-2004`

compacts its atmosphere, while Saturn's is more extended. Nevertheless, during 2004 the Cassini spacecraft revealed oval-shaped anti-hurricanes in Saturn's southern hemisphere. In a similar manner to Jupiter's, Saturn's smaller-scale storm systems are thought to be powered by the motion of gas and heat from the planet's interior. In 2004 a powerful storm, named the Dragon Storm, appeared on Saturn with bright swirling cloud features. It was imaged in infrared light by the Cassini spacecraft, which also detected a very strong source of radio emissions from the same region. The spacecraft only picked up the radio bursts when the storm was appearing from the night side of Saturn, and the emissions stopped as the storm moved into sunlight. The Dragon Storm on Saturn is believed to be a vast thunderstorm, with the radio static resulting from high-voltage lightning strikes. The storm is now thought to be long-lived, feeding on an upsurge of energy from deep within Saturn's atmosphere. It is also interesting that smaller, dark storm features previously noted by the Cassini spacecraft over the same region of the planet subsequently stretched and merged. In a similar fashion to Jupiter's Great Red Spot, therefore, it seems that the little storms plus the thunder help maintain the most powerful weather events on Saturn.

Uranus seems to lack any significant internal heat source. Neptune, however, radiates almost three times more heat than it receives from the Sun. While the cause of this extra heat in Neptune is still uncertain, it is most likely to be energy left over from its formation (as is the case for Jupiter). This stark difference in internal heat sources is matched by a clear difference in the appearance of the planets in terms of weather systems. Neptune displays a more dynamic atmosphere, with several oval storm systems. The largest of these was discovered by the Voyager 2 spacecraft in 1989 and was named the Great Dark Spot. Just like Jupiter's Great Red Spot, Neptune's striking storm was located close to the equator and rotated counter-clockwise. This giant anti-cyclone hosted wind speeds of 1100km/h (684mph). Unlike Jupiter's centuries-old system, however, the

This false-colour image of Saturn reveals a feature, just right of centre, dubbed the Dragon Storm. It resides in a region of the planet's southern hemisphere known for prolific storm activity, and is thought to be a massive thunderstorm that unleashes bolts of electricity 1000 times stronger than lightning on Earth. The storm, which is about as wide as the USA, emits powerful bursts of radio waves, which were detected by the Cassini spacecraft.

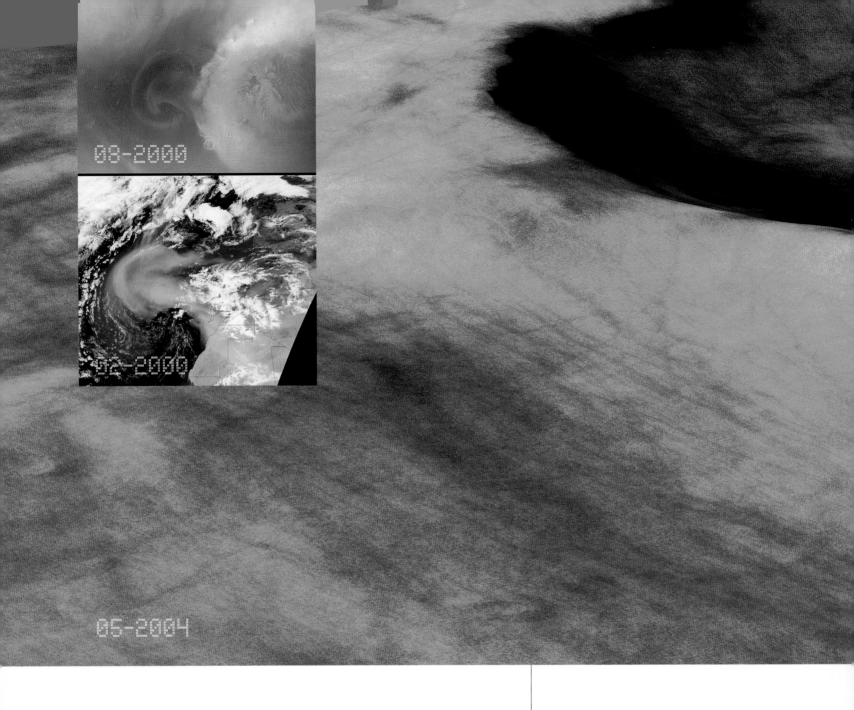

08-2000

02-2000

05-2004

The main image shows a view near a crater in the southern highlands of Mars. The smooth surface, composed of layers of dust and volcanic ash, is starkly marked by numerous dark tracks. These are the scars caused by the movement of dust devils across the Martian surface. The European Space Agency's Mars Express spacecraft has studied several examples of the actions of these mini-tornadoes as they lift away the upper layers of dust to reveal the darker terrain below. The insets show an intriguing comparison between a dust storm on Mars (top) and a storm seen near the coast of northwest Africa on Earth (bottom). On both planets, dust storms can cause environmental changes and control weather patterns.

Great Dark Spot on Neptune had disappeared by the mid-1990s, when the Hubble Space Telescope found no trace of it (though several new, smaller bright spots have appeared).

There she blows!

The giant planets Jupiter and Saturn have a banded appearance, with the former especially boasting bright zones and dark belts. These planet-girdling features trace zones of high and low pressure. They represent a global and fairly stable pattern of east-to-west wind flows. The gas planets also have equatorial flows reminiscent of jet streams on Earth. Jupiter's eastward flowing equatorial jet stream blows at a speed of 360km/h (224mph), but even more remarkable are the 1600km/h (995mph) jets tearing across Saturn; somewhat enigmatically, these speeds are faster than the rotation of the planet itself. Tremendous gusts exceeding 1000km/h (620mph) are also common on Neptune. On Earth, the winds are caused by regions of our planet being warmed at different rates by the Sun. The variations in temperature lead to pressure differences and air moves (as wind) from high pressure to low pressure. The giant planets are, however, much farther away from the Sun and get only a fraction of the solar energy that the Earth receives. Scientists are therefore still grappling with the fact that the gas planets are so cold yet have such active and turbulent winds.

Martian dust storms and devils

One of the unpredictable weather conditions that we on Earth have in common with the red planet Mars is dust storms. On Earth they appear as vast swathes of dust, usually in deserts and dry terrain where fine soil is abundant and easily carried by winds. Great dust storms on Earth include those that swamped the American mid-west in the 1930s and parts of Australia during the extreme drought of 1983. Mars offers scientists a fascinating comparison for these storms, which is important in advance of any future missions to land astronauts on the planet.

Though the atmosphere of Mars is meagre compared to Earth's, it is still sufficiently dense to generate winds and powerful, long-lasting storms. The relatively thin atmosphere and lack of oceans makes Mars' atmosphere very sensitive to changes in temperature. This couples with the planet's non-circular path around the Sun to create uneven heating between the Martian north and south poles over the course of its year. The difference in sunlight causes temperature fluctuations that

control winds on Mars' surface and thus lift dust off its dry landscape. Because vast spans of Mars are covered in fine dust, 160km/h (100mph) winds kick off substantial dust storms. The phenomenon is generally worse in the summer-time southern hemisphere, when Mars is closest to the Sun and basks in its maximum amount of solar heat. One of the largest dust storms on Mars in the past 30 years erupted during late June 2001. It started as a small cloud of dust inside an impact crater on the southern half of the planet. From a modest beginning, it rapidly grew to engulf the entire planet. This dust storm did not weaken for more than three months and was so vast that it could even be seen by astronomers on Earth using modest telescopes. No one is really sure how dust storms on Mars grow to such global proportions. Meteorologists puzzle over how the dust is raised, held aloft, and then accelerated over great distances. In a remarkable example of rapid global warming, the dust storm of 2001 temporarily caused daytime temperatures on the ground to cool by almost 3°C (37°F), while the night-time average increased by 1°C (34°F). This was because the high-altitude dust reflected sunlight and prevented the ground from warming up. At night, however, the dust provided a warm blanket around the planet. These engulfing Martian dust storms are equivalent to shrouding every land continent on Earth in dust at the same time. Dust storms are generally smaller on Earth because our planet is not a global desert like Mars. Regions such as the Pacific Ocean, for example, act to dampen down storms that are originally launched from the Gobi desert.

There are other dusty hazards on Mars. Occasionally tornadoes form that can tower up to 8km (5 miles) high. Known as dust devils, these structures occur when the ground heats up and warm air starts to rise. The soaring air spins and moves across the dry surface, scooping up the abundant loose dust. The rapidly turning features quickly take on a tornado-like appearance. As dust devils travel across the Martian surface, they lift away the upper layers to reveal the darker terrain beneath. Orbiting spacecraft, such as Mars Express, have imaged the dark streaks created by the vortexes set in motion. Aside from the meteorological value of studying dust storms on Mars and comparing them with Earth's related phenomena, the great Martian storms could pose serious threats to robotic and astronaut missions to the red planet. The fast-travelling dust particles could easily cause mechanical failures in instruments, vehicles, and even spacesuits.

Electricity and Radiation

For some time now scientists have debated whether the Earth-sized, planet Venus has a turbulent atmosphere, riddled with lightning strikes. In 1990 the Galileo spacecraft made a tour past Venus en route to its destination, Jupiter. Radio observations made during the flyby revealed small electrical impulses that have been interpreted as evidence of lightning on Venus. The data was, however, acquired from some considerable distance away from the planet, and contrasts with negative results from Earth since then. During its journey to Saturn, the Cassini spacecraft also made two flybys of Venus, the first in April 1998 and the second in June 1999. During the flybys, 9m (30ft) antennae on Cassini were used to "listen" out for high-frequency radio waves that would indicate the presence of lightning in the Venusian atmosphere. The radio emissions are akin to the static heard on radios during thunderstorms here on Earth. The Cassini instrument failed to detect the tell-tale high-frequency signals. It remains a possibility that lower-energy activity may exist on Venus but that it could not travel through the planet's thick atmosphere.

The meteorological conditions in the atmosphere of Venus, plus ingredients such as sulphuric acid, are very different from those of Earth. It would not be surprising then that the electrical activity on Venus may be radically different from lightning on Earth. Studies of lightning on terrestrial planets, such as Venus, are important because frequent electrical strikes can promote the formation of biologically significant molecules. They can also modify the chemistry of the atmosphere, and can pose a threat to space probes sent from Earth.

Fresh perspectives on Venus are offered by the launch on 10 November 2005 of the European Space Agency's Venus Express spacecraft. After a five-month journey, this exciting new mission to the cloud-shrouded planet aims to undertake a detailed investigation of the activity and chemical make-up of its atmosphere. This will also be the first time that a spacecraft has directly probed Venus' blanket of cloud cover to view its surface, by uniquely exploiting observations taken in the infrared regions of the electromagnetic spectrum. The mission is planned to last at least a year and a half, and may be extended depending on the health of the spacecraft. Venus Express will have to withstand far more extreme conditions than orbiters sent to Mars. Its power-generating solar arrays need to be protected from temperatures nearing 250°C (480°F), which includes heat reflected from Venus' atmosphere. We stand to learn a great deal about the dynamic, possibly electric, processes in this hostile environment, including how it interacts not only with the planet's surface but also with material ejected from the Sun.

This full image of the Sun is in fact a mosaic of 36 separate exposures from NASA's Transition Region and Corona Explorer (TRACE) spacecraft telescope. The telescope's specially coated mirrors deliver superb quality images in ultraviolet light that enable astronomers to monitor explosive changes on the Sun over time periods of minutes. Violent eruptive events are seen, such as solar flares that can blast million-degree hot gas into inter-planetary space.

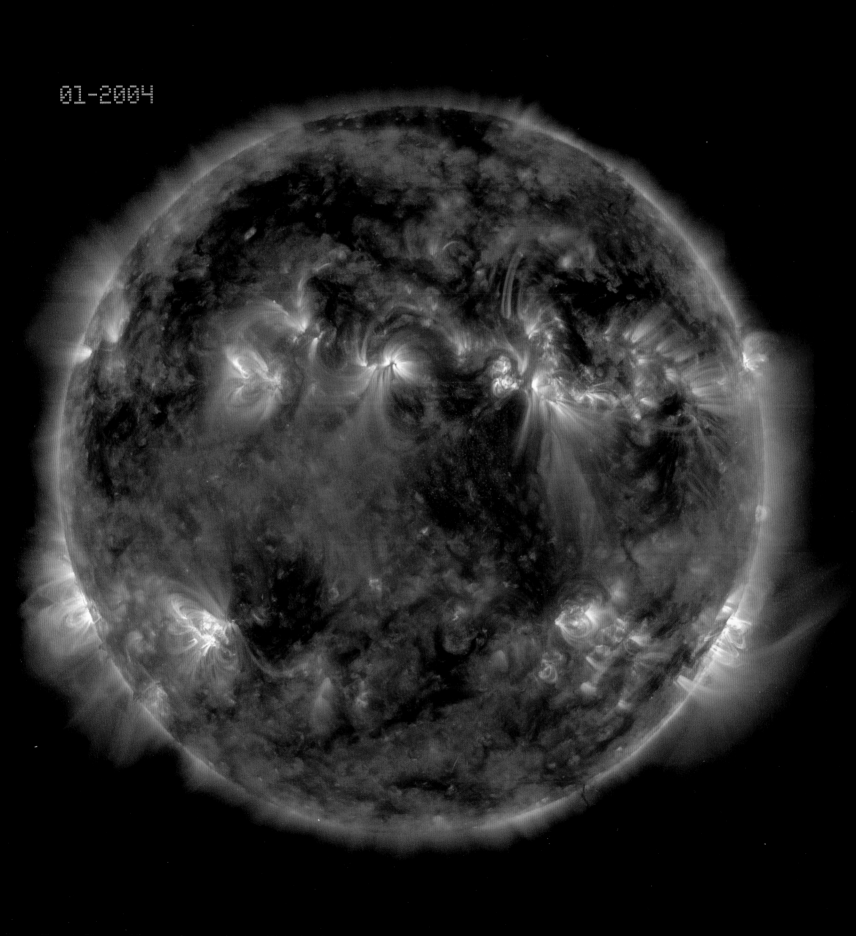
01-2004

Eruptions from the Sun

With its dominant mass and vast energy output, the Sun is the centrepiece of our solar system. This great ball of mostly hydrogen and helium gas is our back-yard star; a very ordinary one, in fact, compared with some of the most luminous stars that inhabit our Milky Way Galaxy. Powered by a thermonuclear fusion reactor in its 15 million °C (60 million °F) hot core, the Sun bathes the planetary system with its radiant energy. It illuminates and warms the planets, and we have already seen that the solar heat is pivotal in driving much of the weather in the atmospheres of both the terrestrial and the gas planets.

The Sun itself is, however, also a stormy beast, with outer layers of seething hot gas that can erupt in phenomenal manners. The biggest and most powerful explosions in the solar system occur on the violent Sun and come in two principal forms, known as solar flares and coronal mass ejections. The solar flares are brief and catastrophic events that flash across regions of the Sun. In barely a few minutes they release enormous amounts of energy, flooding the solar system with a burst of radiation and charged particles, such as electrons and protons. The flares erupt from very active regions of the Sun where the magnetic field is especially strong. Typically, these spectacular solar storms can each release energy equivalent to that of 100 megaton nuclear bombs. When the Sun is at the peak of its activity, flares can erupt daily, or even more frequently. One of the largest flares recorded in recent times occurred on 28 October 2003. Billions of tons of solar material were ejected into interplanetary space. The event was followed by a series of solar storms that slammed into the Earth during early November 2003. The most energetic of these events produced intense solar radiation that resulted in several days of radio blackouts in the northernmost latitudes. In September 2005, another particularly large flare erupted on the Sun, resulting in widespread blackouts of high-frequency radio communications.

Large flares on the Sun are sometimes followed by the other explosive activity, the coronal mass ejection (CME). Enormous bubbles of electrified gas are spewed out from the Sun into space over a period of several hours. Depending on how stormy the Sun is, CMEs can occur several times per week. They can occur anywhere on the Sun and provide a means for our star to release an excessive build-up of magnetic energy. When CMEs head straight for the Earth, they can cause geomagnetic storms on our planet and represent a major hazard to satellites

and astronauts. Similar to the threat from solar flares, the turbulent waves of charged gas from the Sun can damage sensitive communications electronics of satellites and interfere with radio and television signals. When a CME erupts and is directed toward Earth, it can take two to three days for the radiation and plasma to reach our planet. That is normally enough notice to cancel spacewalks on the Space Shuttle or International Space Station and temporarily switch off the most sensitive instruments on satellites. Storms in space also have potentially serious consequences for the safety of satellites placed in low-Earth orbits, barely a few hundred kilometres

above the planet's surface. During increased periods of solar activity, the extra energy generated can puff up the Earth's atmosphere and thus increase the friction on low-orbiting satellites. This phenomenon caused the early demise of the Solar Maximum Mission in 1990. Periodically, the International Space Station also has to have its orbit boosted to avoid a similar fate. For this reason, space weather forecasting has become a very serious matter. Many defence agencies and governments acknowledge Earth's vulnerability and have established dedicated facilities to monitor solar activity, such as the Space Environment Center in the USA.

Magnetic shields

Fortunately for us, we are protected from the harmful effects of powerful solar storms by an invisible magnetic shell around the Earth known as the magnetosphere. (Of course the Earth's atmosphere also guards against some of the Sun's radiation.) The Earth's magnetosphere was discovered in 1958 using satellites built to detect high-energy charged particles. You can think of the magnetosphere as the region around Earth that is affected by our planet's magnetic field. Particles ejected from the Sun speed toward Earth at about 450km per second (280 miles per second). Near our planet, however, these energetic and electrically charged emissions are deflected by the Earth's magnetic field.

The shape of the cavity traced by the magnetosphere is somewhat like a wind sock pointing away from the Sun. It is greatly distorted by the flow of material from the Sun, which drags it "downstream", making a long tail that extends far into space. The Earth's magnetosphere is continually bombarded by high-speed charged particles from the Sun that would be extremely harmful to the human body. Some scientists even

The launch of a powerful eruption is caught in this image from the Extreme Ultraviolet Imaging Telescope. The glare from the Sun is blocked by an occulting disc and the surface of the Sun is marked by the white circle. The spectacular explosion, known as a coronal mass ejection, blasts billions of tons of solar material at more than 3 million km/h (1.9 million mph). These events last a few hours and are thought to be linked to changes in the Sun's magnetic field.

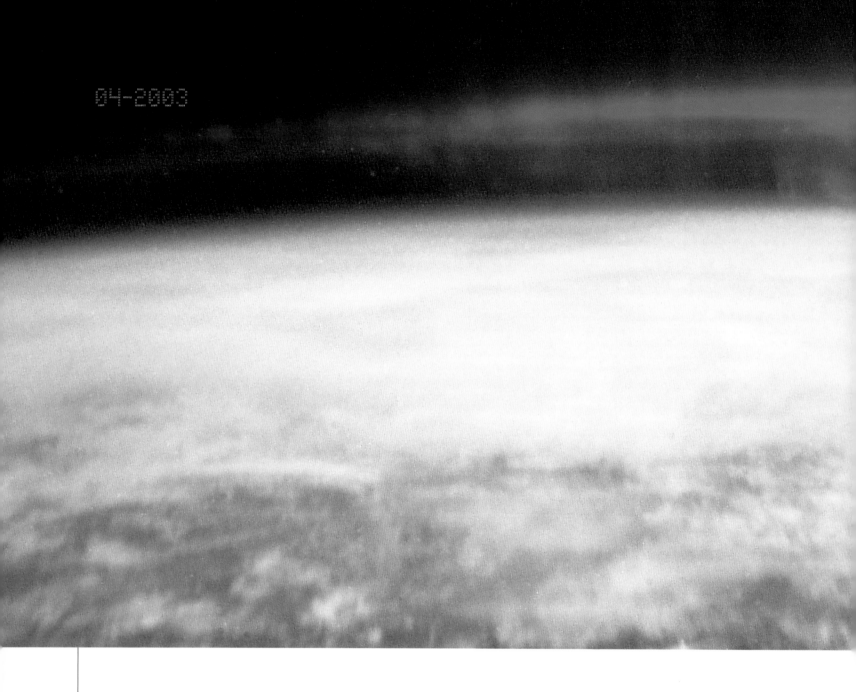

The magnificent sight of a green aurora in the Earth's atmosphere as pictured by astronauts on board the International Space Station. These dynamic electrical phenomena generally occur at an altitude above 100km (60 miles) and can occasionally be seen three times higher up, at about the same level as the orbit of the Space Station. The auroral displays do not, however, pose any threat to the spacecraft.

believe that complex life forms might not have started on Earth if it had not had the protection of this magnetic shield.

Many other planets and moons in the solar system also have magnetic fields and thus magnetospheres. Jupiter's magnetosphere is the most impressive, with a long tail extending away from the Sun all the way to Saturn's orbit. That makes Jupiter's magnetosphere far larger than the Sun and the largest single object in the solar system. If Jupiter's magnetosphere glowed in visible light, it would appear several times larger than the full Moon in our night skies. This enormous structure tells us that

Jupiter's magnetic field strength is nearly 20,000 times greater than Earth's. Saturn's magnetic field is about 1000 times stronger than our planet's, which is sufficient to create a magnetosphere that is large enough to contain the beautiful rings and at least sixteen of Saturn's innermost moons. The terrestrial planets Mars and Venus have considerably weaker magnetic fields than Earth, because either they don't have molten cores or they spin too slowly. Because of this, electrical currents, which are a pre-requisite for creating strong magnetic fields, are not generated within these two planets.

Spectacular light shows

Storms from the Sun that flood the solar system with highly charged particles can dump vast amounts of energy into the planets' magnetospheres, sometimes with fantastic consequences. Among the most remarkable phenomena are the aurorae, often referred to on Earth as the northern and southern lights. Named after the Roman goddess of dawn, aurorae are one of the most beautiful natural displays on our planet.

The charged plasma from the stormy Sun can penetrate deep into the Earth's magnetosphere near the North and South Poles. These are the regions where the magnetic field is the weakest and the "shield" is partially open. Electrical currents are created and as the particles enter the upper atmosphere they collide with atoms of oxygen and nitrogen. This happens 150–400km (93–250 miles) above ground level, causing the oxygen and nitrogen to glow. The results, seen at polar latitudes, are dazzling displays of green, blue, red, and white lights in the night sky. They mostly appear as wispy curtains, shimmering with pale rays. In general, the closer to the North (or South) Pole you go, the more likely you are to see aurorae.

Aurorae have also been detected on all four of the giant gas planets, with those on Jupiter being the most extensively studied. The latest spacecraft missions, such as Galileo and Cassini, have combined with the Hubble Space Telescope to provide fresh perspectives on the intriguing nature of aurorae on Jupiter. While we understand well how ejecta from storms on the Sun control the Earth's aurora, this interaction is considerably more complex on Jupiter. The aurorae on Jupiter are 1000 times more energetic and are produced by two different methods. The first is similar to that which occurs on Earth and is the consequence of the flow of energetic solar particles that are trapped by the enormous Jovian magnetosphere. Another remarkable source of material, however, is from active volcanoes on Jupiter's moon Io (see page 63). The eruptions on this moon provide a huge extra reservoir of electrons and ions that become trapped inside Jupiter's magnetosphere. These charged particles are then accelerated down into the atmosphere above the north and south poles of the planet. The particles collide with the gasses to glow and produce glowing aurorae that are continually active on the planet. The aurorae have been imaged in ultraviolet, infrared, and visible wave-bands. Equally fascinating were the X-ray images recorded by NASA's Chandra satellite during 2003, which revealed the presence of highly charged particles slamming into Jupiter's atmosphere. The X-ray observations suggest that 10 million volts of electricity are generated in Jupiter's aurorae. This means that these aurorae are more than 100 times stronger than the lightning bolts seen on Earth in a thunderstorm.

Beautiful aurorae are also found on the gas planet Saturn. Ultraviolet images captured in 2005 by instruments on board the Cassini spacecraft during its mission to study the planet and its rings, have extended earlier discoveries made by NASA's Hubble Space Telescope. The observations reveal oval-shaped glowing emissions around Saturn's poles. The lights are caused by hydrogen gas being bombarded by electrons. While Saturn's aurorae mostly appear as luminous rings of energy, occasionally

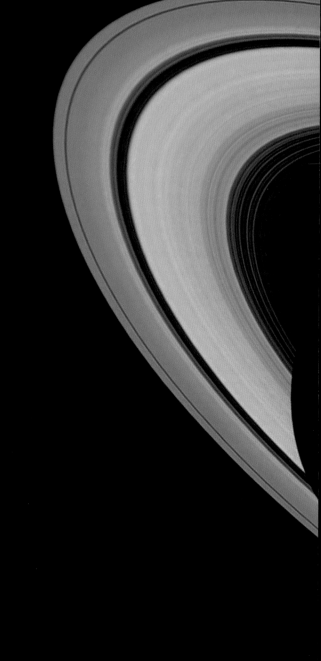

The tell-tale ring of auroral light circling the southern polar region of Saturn. With a typical lifetime of several days, Saturn's aurorae are brighter and more powerful than those of Earth. The example shown here is from a series of ultraviolet images taken by the Hubble Space Telescope. The aurorae of Saturn change from day to day, in direct response to the streams of charged particles coming from the Sun.

03-2004

more enigmatic spiral-shaped patterns are observed. The Hubble Space Telescope images show that Saturn's aurorae react rapidly to changes in the material that is streaming away from the Sun. These remarkable views provide us with evidence that Saturn's aurorae vary in detail from day to day, very much like those on Earth. However, the big difference is that the aurorae light shows on Earth typically last only about 10 minutes, whereas the displays on the giant planet Saturn can persist for days.

Equally intriguing are light emissions detected on the planets Venus and Mars that may be interpreted as aurorae. Their presence on these terrestrial planets is surprising because these bodies have essentially no magnetic fields and magnetospheres to trap and direct particles from the Sun. During 2004, the

A stunning aurora glows eerily around the northern polar region of Jupiter. The image was taken in ultraviolet light with the Hubble Space Telescope and reveals a complex pattern of lights resulting from high-speed electrons entering the giant planet's upper atmosphere.

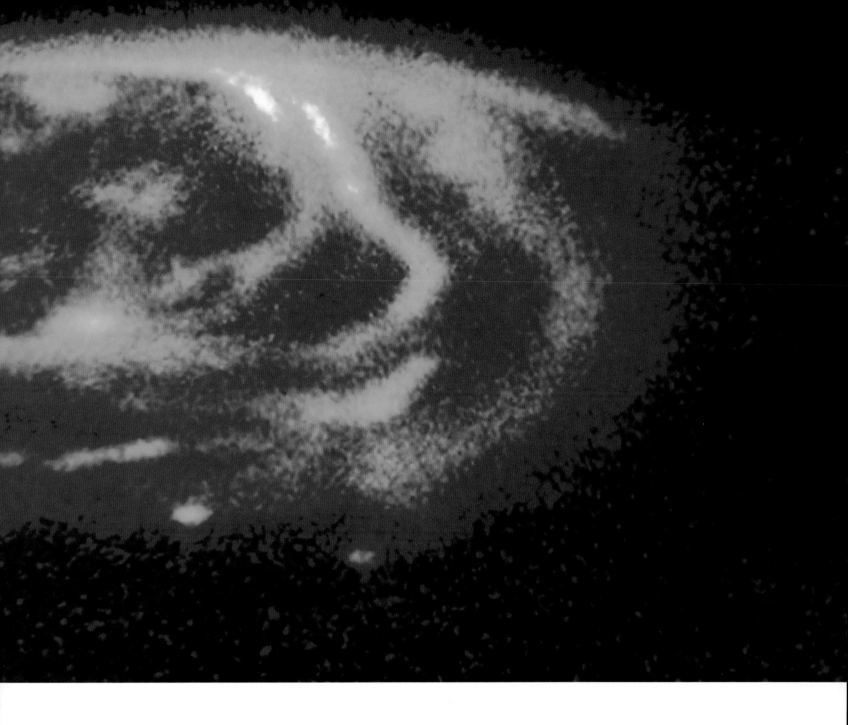

European Space Agency's Mars Express spacecraft made the first detection of aurorae on Mars, but this is a unique phenomenon unlike aurorae seen on Earth, Jupiter, and the rest of the gas planets. The data from Mars Express suggest that the magnetic fields of the planet are linked with the rocky surface crust. Electrons then speed along the crust and charge up the atmosphere of Mars to create auroral lights.

It has now become clear that spectacular auroral light displays can be seen right across the entire solar system. Aside from the sheer beauty of these phenomena, the variety of their forms and properties provides scientists with some fascinating insights into the magnetic nature of every one of the host planets in our solar system, as well a greater understanding of the widespread influences of a stormy Sun.

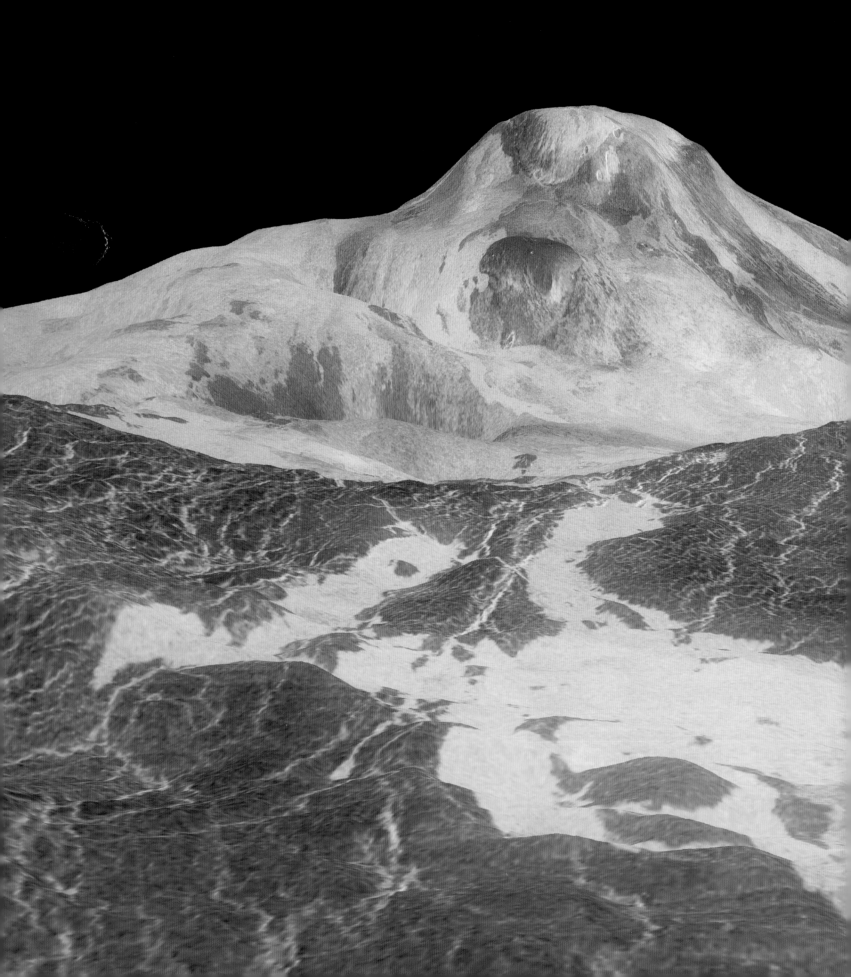

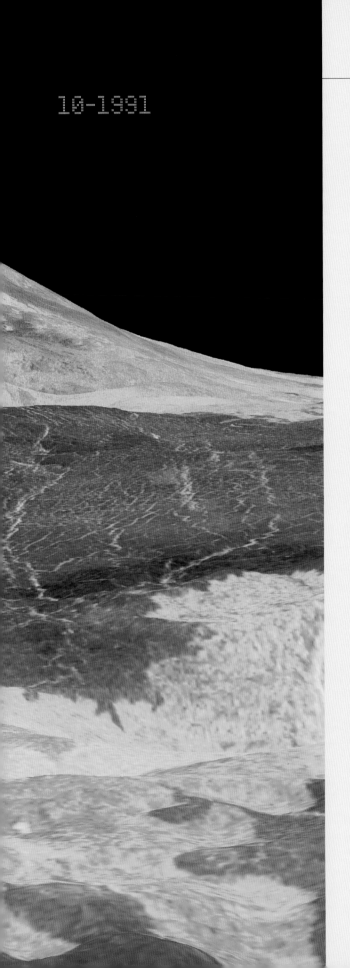

10-1991

A perspective view of Maat Mons, which is a 8km (5 mile) high shield volcano on the planet Venus. The image is based on radar mapping conducted by NASA's Magellan spacecraft, using hues simulated from colour images previously obtained by the Venera 13 and 14 landers of the former USSR. Ancient lava flows can be seen extending for hundreds of kilometres.

Volcanoes – Alive and Ancient

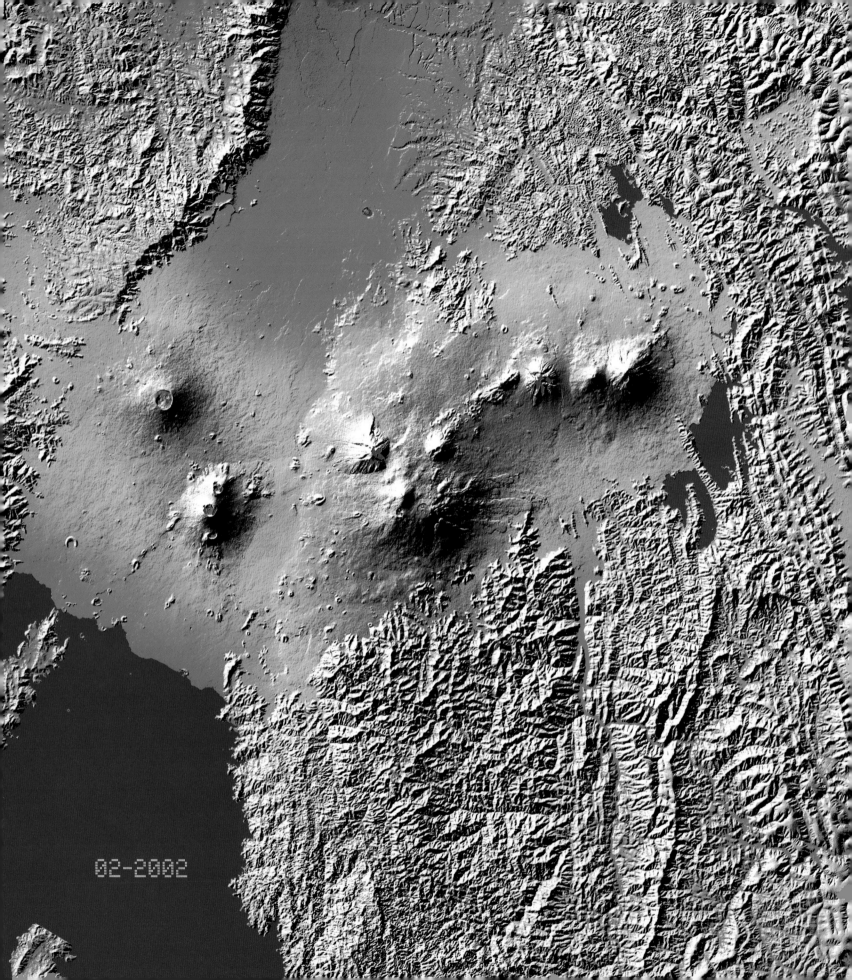

02-2002

Radar antennas flown on board the Space Shuttle Endeavour have been used to acquire very detailed data on the elevation and relief of the Earth's terrain. The radar signals are combined to produce three-dimensional images, resulting in the most complete topographic map of the Earth's surface ever obtained. The example shown here reveals features 30m (98ft) across, in a 110km (69 mile) square region dominated by the volcano chain of the East African Rift, bordering Congo, Rwanda, and Uganda.

Volcanic eruptions on Earth rate as one of nature's most spectacular and frightening displays of power. They represent a fundamental geological process for sculpting the surfaces of planets and active volcanoes play a significant role in building and shaping the landscape. Earth is not the only planet, however, where volcanic activity has transformed the surface. This chapter explores what the latest spacecraft probes have revealed about the action of volcanoes in other parts of the solar system. These missions have beamed back images of remarkable structures that dwarf any that we find on Earth. Several hundred volcanoes have been detected on Venus and, while most of these are long extinct, some may be active today. The planet Mars boasts the most gargantuan volcanoes, which tower tens of miles high over the Martian landscape. A true wonder of the solar system is Io, the innermost moon of the giant planet Jupiter. Io is the most currently active volcanic body known. Its prolific eruptions can resurface the entire orb every million years. Saturn's smooth moon called Enceladus is equally fascinating, where ice volcanism can launch plumes of water vapour. This chapter brings together these exotic sites of volcanic activity in our solar system. Together, they represent fantastic comparative laboratories for understanding not only our own dynamic planet, but also some truly remarkable worlds beyond it.

Volcanoes and Life

Mercury, Venus, Earth (plus its Moon), and Mars are terrestrial planets, with solid surfaces that preserve a record of how each planet has evolved over billions of years. Volcanic eruptions play a major role in creating new terrain and partially obliterating existing landscape. Following their formation almost 4.5 billion years ago, the first 500–700 million years of the terrestrial planets were extremely violent. Their geological activity levels were very high, with heavy bombardment by asteroid-sized bodies and rampant volcanic activity. Billions of years later, the planets are now middle-aged and their heat engines, which power their volcanoes, are not what they used to be. Different planets have different cooling-off histories.

It is perhaps telling to note that today our Earth is not only the most biologically alive planet but also the most geologically active. Although there are almost 1500 named volcanoes on Earth, on average about 20 volcanic eruptions occur at any given time (most of these are underwater). Catastrophic eruptions that result in immediate changes to the landscape are fairly rare. The eruption of Krakatau in 1883 between the islands of Java and Sumatra still rates as one of the largest explosions on Earth in recorded history. It destroyed much of the Krakatau island, with only a relatively small remnant surviving. The most voluminous pouring of lava in recent times has come from the Kilauea volcano in Hawaii. The flows since 1983 have added almost 200 hectares (495 acres) to the island's shore. The eruption of Mount St Helens in 1980 was another example of relatively instantaneous changes to the terrain. The avalanche of debris turned more than 515 sq km (200 sq miles) of rich forest into a lifeless landscape. It produced enough material to resurface Washington DC to a depth of 4m (14ft). The grandest volcano-driven landscapes, however, take shape over hundreds of thousands of years.

Ironically, despite the destructive and threatening nature of volcanoes, they are in fact a vital component of life on Earth. Besides creating new land masses, volcanic eruptions release vast supplies of carbon dioxide, nitrogen, and water into the Earth's atmosphere. Our planet did not form with its atmosphere in place. Key constituents such as water, methane, and carbon dioxide were originally trapped in rocks in the interior of the Earth. It was only when the rock melted that the gas was released to the surface. This process of "outgassing" probably produced most of the condensed water vapour that formed the oceans, and provided gases for the atmosphere. Volcanic eruptions are the main source of this outgassing, which even today produces water, nitrogen, and sulphur. Volcanic ash is also an excellent plant fertilizer. It seems that in Earth's earliest history the heat from volcanoes may have created fixed forms of nitrogen for use by primitive micro-organisms. Living organisms need nitrogen to survive. Today, bacteria and fungi play this role

The eruption of Mount Etna on the island of Sicily is dramatically captured in this image taken by NASA's Terra satellite. Mount Etna is Europe's most active volcano and occasionally exhibits spectacular plumes and fountains of lava from its summit and flanks. Satellite observations of the Earth's volcanoes play a vital role in improving our understanding of the effects of eruptions on our global climate.

11-2002

of converting nitrogen into a form that plants can use, which in turn is absorbed by animals along the food chain. Volcanoes on the Earth's sea floors also play an important role in the chemical balance of ocean waters and can accumulate huge amounts of metals such as copper, lead, and zinc.

Anatomy of a volcano

So what causes volcanoes on Earth? Our planet is made up of three principal layers called the crust, mantle, and core. The crust is the thin outermost skin of the Earth and is barely 6km (4 miles) thick under the oceans and 50km (30 miles) on the continents. The Earth's thickest layer is the rocky mantle, below the crust. Extending 3000km (1865 miles), the mantle is mostly solid, but also extremely hot. Some of this material melts in the heat to form magma. Digging to about 3000km (1865 miles) below the planet's surface we would arrive at the the Earth's core. First is the liquid outer layer, which is followed by an inner core that is mostly dense solid iron at temperatures of around 5500°C (9930°F). That makes the Earth's centre almost as hot as the Sun's surface. A trip from the surface of the Earth to its centre would be the equivalent of climbing up Mount Everest almost 720 times.

To understand volcanoes we need to look at the Earth's crust in a little more detail. It turns out that the crust is not a

A radar-based image of Mount Saint Helens in Washington State, USA, constructed from the topography experiment flown on the Space Shuttle Endeavour. Mount Saint Helens is an example of how a stratovolcano can shape the surface of a terrestrial planet on relatively short timescales. Its eruption on 18 May 1980 resulted in the collapse of the upper 400m (1300ft), which slid over the surrounding terrain to leave a shorter, crater-topped mountain. Since 1980, 15 cases of volcanic unrest have been recorded in 33 volcanoes in the USA that are being monitored.

single continuous layer around the Earth; it is broken into seven large, snugly fitting plates, plus several smaller ones. These plates are constantly moving, driven by the motion of magma in the mantle below. When the magma moves, the plates of the crust shift and occasionally they run against each other. Alternatively one plate might get pushed below another, or two plates pull apart. These powerful plate motions cause earthquakes and volcanoes. Any natural break in the Earth's crust, where melted rock, ash, gases, and steam appear, is a volcano. The hot magma and gases are forced up to the surface, forming the characteristic volcanic mountain shapes. Volcanoes on Earth can become active over different periods of time and they do not always erupt in violence; they can seep lava in a slow placid manner.

The most spectacular volcanoes usually erupt in areas on the crust known as hot-spots. These are weak regions in the crust that give way to the immense pressure of the hot magma beneath. The Hawaiian islands were created by such a hot-spot some 70 million years ago. The most prolific volcanoes on Earth can have global effects, such as the June 1991 eruption of Mount Pinatubo in the Philippines, which spewed out 20 million tons of sulphuric acid that encircled the Earth in three weeks. The expulsion of vast amounts of ash and gases influenced the climate of the world for several years, with an average drop in global temperatures of almost 0.5°C (0.9°F).

Awe-inspiring volcanoes are not continually erupting. They can lie quiet and dormant for thousands of years, but they always have the potential to come to life again. Volcanoes on our planet that have not erupted for more than 10,000 years are usually labelled extinct, because they have probably exhausted any supplies of magma. People living close to active volcanoes need early warnings of potential eruptions. Scientists therefore monitor volcanoes to understand what triggers them so that they can alert the nearby communities. They study the motion of magma below a volcano or look for evidence for seismic activity, such as earthquakes. These events, plus any accounts of volcanic gases being released from vents, help scientists to predict when a volcano is about to erupt.

Volcano allsorts

Volcanoes don't all have the same eruptive behaviour, they also do not all have the same form or structure. A variety of volcano types and shapes is seen on Earth, just as on other bodies in the

solar system. All volcanoes will have a summit crater from which the lava flows, but the surrounding structure can vary widely in shape. The form of these edifices is determined not only by the properties of the volcanic material but also by the nature of the volcanic eruptions themselves. There are three main volcanic shapes, known as stratovolcanoes, shield volcanoes, and caldera volcanoes.

About 60 percent of Earth's volcanoes are stratovolcanoes. They have tall, symmetric, and steep mountains, with viscous lavas that allow pressure beneath the surface to build to extreme levels. The build-up of pressure leads to highly explosive eruptions. Earth's tallest stratovolcanoes are a pair, each over 6700m (21,900ft) above sea level, standing in the northern Andes of Chile. Mount St Helens in the USA is the youngest and most active example of this class of volcano.

Shield volcanoes usually cover the largest areas of land. They form when very fluid (less viscous) lava flows over vast distances, spanning perhaps hundreds of miles. The extensive flow results in a shallow-sloped volcano that resembles a soldier's shield, hence its name. Mauna Loa in Hawaii is the largest shield volcano on Earth. This volcano rises more than 4km (2.5 miles) above the sea level and covers an area of more than 5200 sq km (2000 sq miles). It is also still one of the Earth's most active volcanoes; the most recent of its major eruptions occurred in 1984. However, there have been more than 30 other eruptions of the Mauna Loa volcano over the past 160 years.

Sometimes the entire top of a volcano blows off during a major eruption. If huge amounts of magma are lost from within the volcano, an extensive empty cavern can form. Known as a caldera, this crater-like structure is really a vast chamber drained of magma. One of the largest examples on Earth is the Aniakchak caldera in Alaska. With a diameter of 10km (6.2 miles) and a depth of 500–1000m (1640–3280ft), it is part of the Aleutian mountains and last erupted in 1911. Ancient calderas on Earth are sometimes filled with water to make circular lakes, such as Crater Lake in Oregon, USA. This was formed after a series of powerful explosions about 6500 years ago. The volcano's top was blown off, together with masses of volcanic ash and dust. These massive events quickly drained the supply of lava, leading to a collapse of the weakened upper regions. The caldera subsequently flooded with water to form the magnificent sight that greets modern visitors.

Ancient Volcanoes of our Neighbours

The range of volcanic phenomena witnessed on the Earth is somewhat dwarfed when compared with the dazzling variety of features on show elsewhere in the solar system. Ancient, long-dormant volcanism is a major feature of the surfaces of the terrestrial, or rocky, planets Venus and Mars, and also Earth's Moon. While powerful monitoring equipment has allowed scientists to advance our understanding of volcanic activity on Earth, comparisons with other planets (known as planetology) offer exciting prospects to study the evolution of the solar system itself.

The Mons of Venus

Volcanoes have played a major role in shaping the surface of Venus. More than three quarters of its surface is covered in lava plains, which are thought to be 500 million years old. This Earth-sized planet has more volcanoes than any other planet in the solar system. More than 1600 prominent volcanic features have been discovered, and there may be in excess of 100,000 smaller volcanoes; we just haven't counted them all up yet. The majority of volcanoes on Venus have been extinct for hundreds of millions of years and none is known for certain to be active today. Despite the efforts of spacecraft missions such as Magellan in the 1990s, our monitoring of Venus is not very extensive and it is possible that some volcanism may still be

ongoing on the planet. Indeed Venus' thick and totally unbroken cloud cover exhibits fairly frequent changes in the levels of sulphur dioxide that are present in the atmosphere, which may be an indication that there are current volcanic eruptions somewhere on the surface.

What the Magellan spacecraft did reveal with its cloud penetrating radar imaging is that Venus has a spectacular variety of volcanoes. The majority are shield volcanoes, and at least 150 large examples have a "shield" between 100km (62 miles) and 600km (373 miles) across. The corresponding heights are 0.3–4km (0.2–2.5 miles). The Venusian shield volcanoes are much broader and with gentler slopes than the great shield volcanoes on Earth. Most also have some form of caldera on the summit, with the slopes covered in extensive ancient lava flows. The volcano named Sif Mons stands about 2km (1.2 miles) tall near the equator of Venus, with a remarkable base almost 300km (186 miles) across. In comparison, the Hawaiian volcano Mauna Loa has a span of only 120km (75 miles). The flows on Sif Mons spread 100km (62 miles) on its shallow gradients, which must mean that the lava was very fluid. Another example of a shield volcano on Venus is Gula Mons, which is 3.2km (2 miles) high and 400km (250 miles) wide and is thought to be a hotspot volcano like those in Hawaii. The largest known volcano on Venus is Theia Mons, which is over 4km (2.5 miles) high and its

A radar-based reconstruction of the once-molten surface of Venus is seen in this detailed image from the Magellan spacecraft. There are circular dome-like hills, typically 25km (15 miles) across, with a maximum height of around 750m (2460ft). The precise volcanic process that resulted in the formation of these domes remains unknown. It is possible to discern here surface features down to a resolution of 120m (400ft).

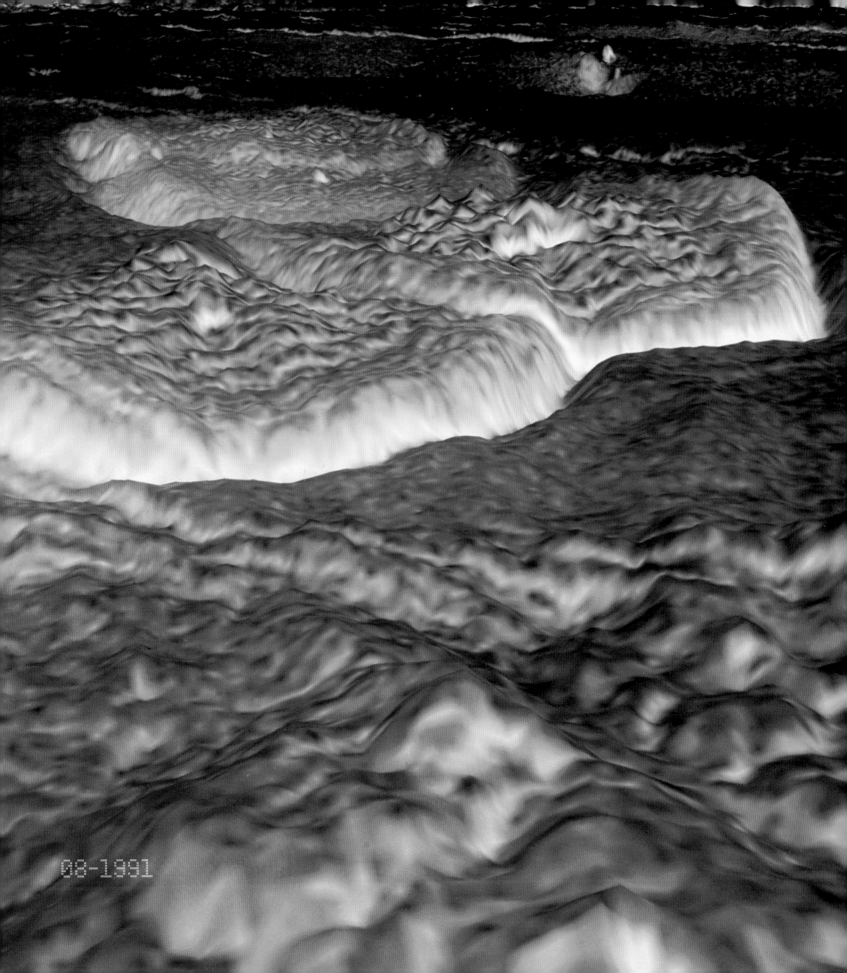
08-1991

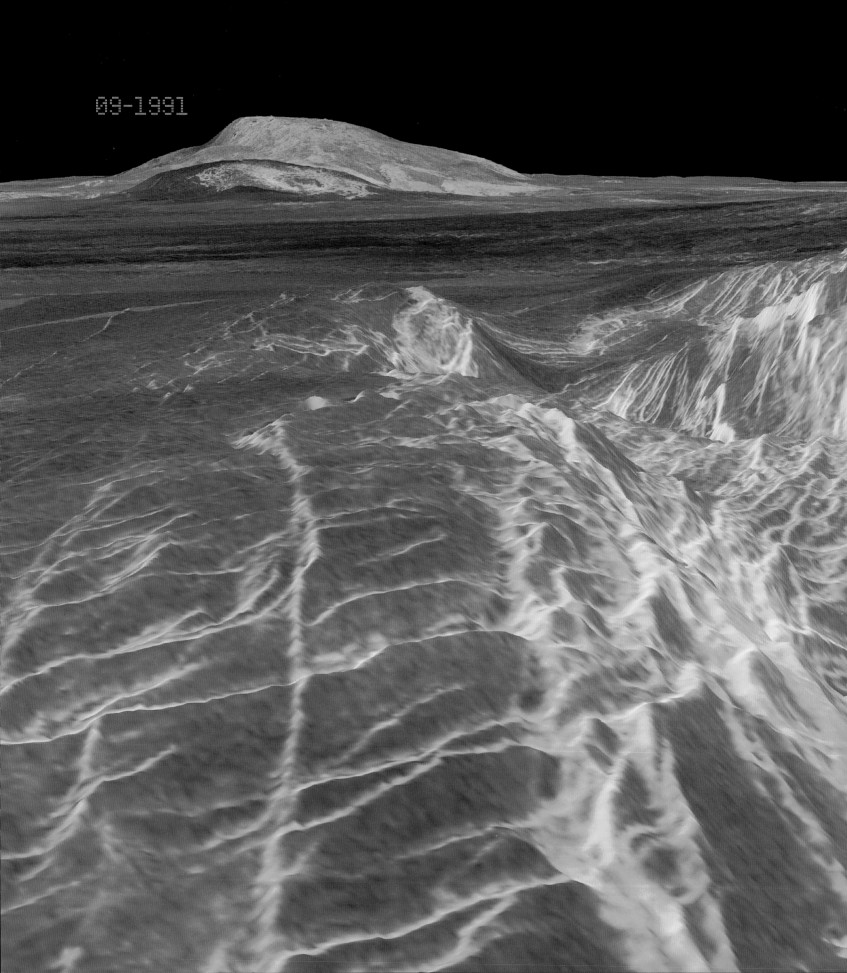

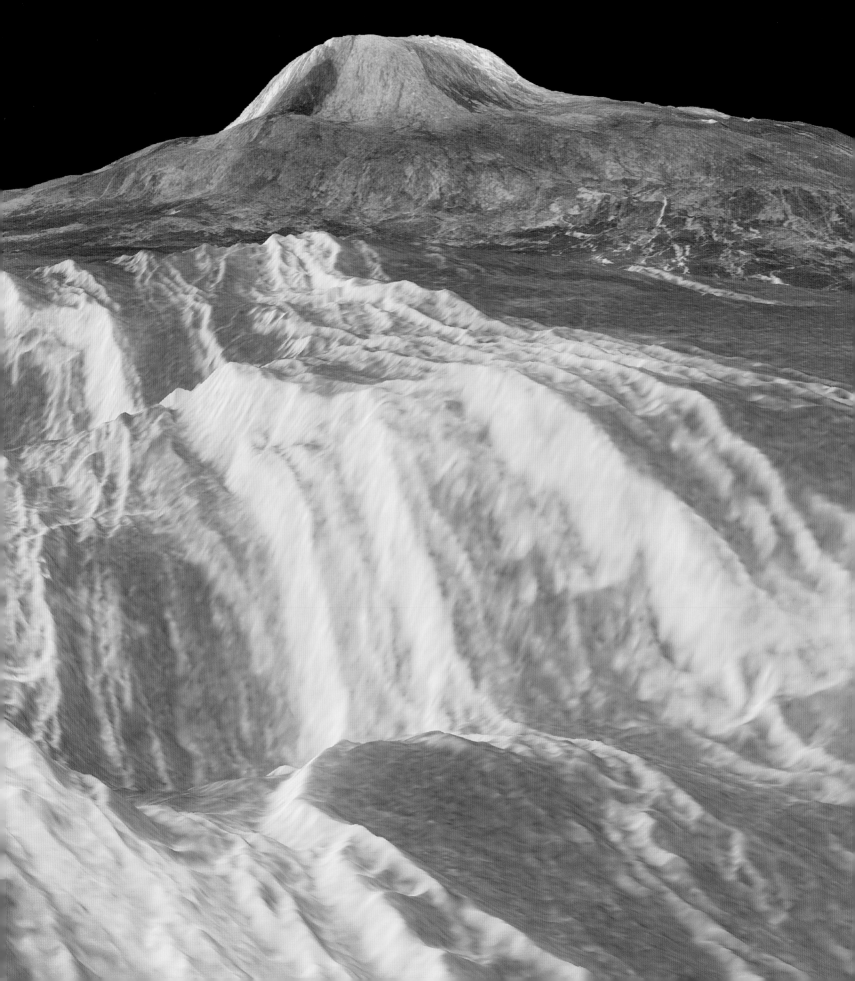

Imagery from the Magellan spacecraft and altimetry data are combined to yield this exquisite view of two massive shield volcanoes on Venus. Gula Mons on the right is 3km (1.8 miles) high, and Sif Mons, left is 2km (1.2 miles) tall with a base diameter of 300km (186 miles). The two volcanoes are about 730km (455 miles) apart. The caldera of Gula Mons is estimated to be more than 100km (62 miles) across. Though these volcanoes are long extinct, recent evidence leaves open the possibility that the planet may still be volcanically active, but at a considerably more modest level.

lava flows spread across an impressive 800km (500 mile) region. Highly fluid lava flows can travel great distances on Venus, and the planet's surface bears the hallmarks of extraordinary lava rivers. Almost 40 ancient channels carved by these molten rivers that are at least 100km (62 miles) long are known. The most remarkable channel, however, is the Hidir lava channel that meanders for 7000km (4350 miles), which is a comparable length to the river Nile on Earth.

The volcanic structures on Venus can come in some unusual forms, such as the circular features called "pancake domes" that are scattered all over this seething planet. These pancake domes can be 45km (28 miles) in diameter and up to 3km (2 miles) high. Intricate patterns of cracks and pits can be seen on the domes. There are similar features on Earth but they are much smaller and never so remarkably symmetrical. The domes on Venus are thought to form from a single one-off eruption, somewhat like a belch of viscous lava.

Another peculiar feature unique to Venus is formed by groups of volcanic faults called coronae; they have are no counterparts on Earth. These mounds are the largest volcanic structures on Venus, typically oval-shaped and hundreds or even thousands of miles across. Almost 400 coronae have been located on Venus. They are usually marked by concentric patterns linked to a sequence of volcanic eruptions and probably formed over hot-spots in the planet's crust, or mantle, resulting in plumes of rising magma that are occasionally turned off before they break through the surface.

Towers of Mars

The red planet Mars has in the distant past experienced substantial periods of volcanic activity. However, it has probably been inactive for the past billion years. Because Mars is smaller than Earth, its interior has cooled a lot more and thus the planet is expected to be less volcanic than Earth. There is some evidence of solidified molten lava that's only 180 million years old, which is young given that the solar system is 4.5 billion years old. Nevertheless, the great Martian volcanoes are unlikely to come to life again. Over the next few billion years Mars is expected to be as geologically inactive as Earth's Moon or the planet Mercury.

Mars does, however, host some of the most spectacular, towering shield volcanoes. One of these is Olympus Mons, which is the largest known volcano in the solar system. It is in fact one of four great volcanoes located in a region known as the Tharsis Bulge, just south of the planet's equator. Olympus Mons is a circular structure, with a base about 600km (373 miles) across. Its peak stands at a spectacular height of 26km (16 miles) above the average Martian terrain, which makes it almost three times as high as Mount Everest on Earth. The volume of this enormous mountain on Mars is 100 times greater than that of the largest volcano on Earth. Olympus Mons alone would span the entire chain of Hawaiian Islands. Like all shield volcanoes on Earth, Olympus Mons has a gentle slope of between one and five degrees over most of the structure. So it would be more of a hike than a climb up the low

This image from NASA's Mars Global Surveyor orbiting mission shows the dormant volcanoes on the (northwest) Tharsis region of Mars. The vast features are lying under the blue-white ice clouds, which make up an overall scene of weather patterns that includes the polar ice caps. Mars has the largest shield volcanoes in the solar system, though most volcanic activity stopped one to two billion years ago. The chance of finding an active volcano on Mars now is slim.

04-1999

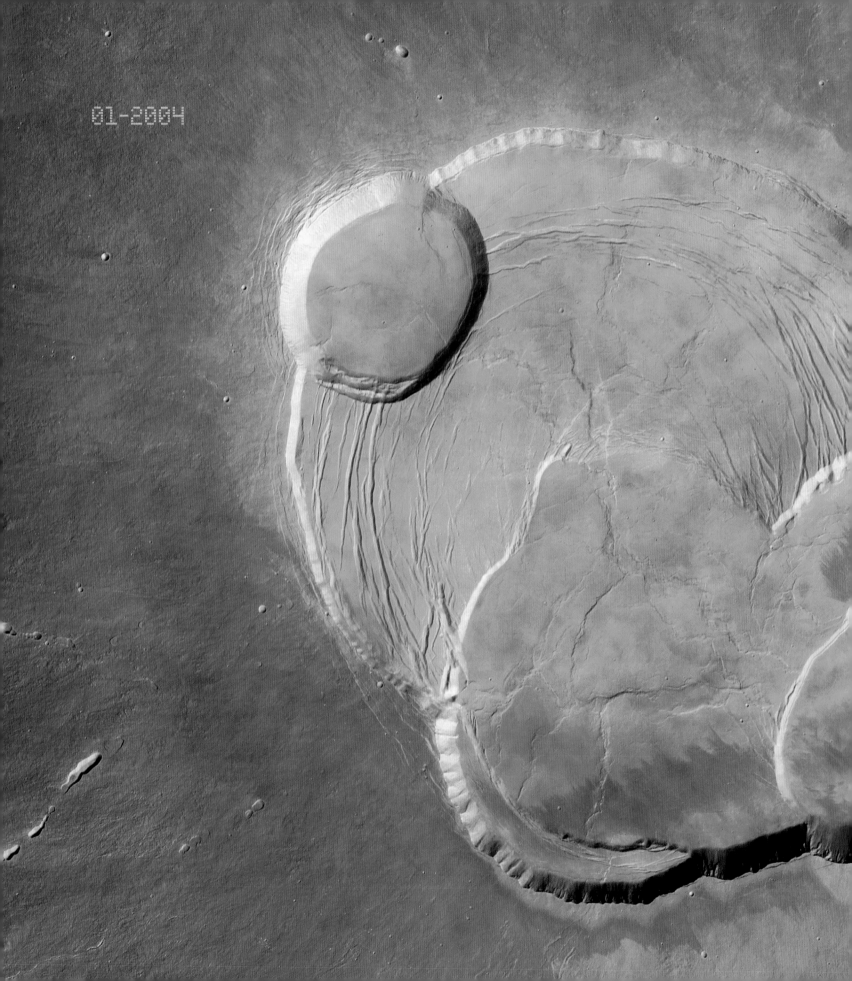
01-2004

slopes, though you would have to contend with a very rugged surface created by overlapping lava flows. It's almost beyond imagination what the prize panoramic view would be after completing such a challenge. The caldera formed at the summit of Olympus Mons was beautifully imaged by the European Space Agency's Mars Express spacecraft during 2004. It formed after lava production had ceased, causing the surface to collapse over the emptied magma chamber below. The caldera is 80km (50 miles) wide, with a depth of 3km (2 miles), and there is some evidence of multiple collapse craters around its circumference.

Olympus Mons probably formed between one and two billion years ago, though some of its youngest lava flows date back less than 200 million years or so. These younger flows were relatively minor and are likely to represent the last spurts of a previously great volcanic outpouring. So how did Olympus Mons get so tall? It was certainly built from a very large magma chamber deep within the Martian crust. The enormous size is probably due to a combination of very frequent eruptions and Mars' low gravity compared with Earth's. Another important reason for the great magnitude of this volcano is that the volcanic hot-spot regions on Mars remained fixed in the mantle for hundreds of millions of years. Unlike Earth's, the crust of Mars is not broken into plates that move around and crash into each other. Because the Martian crust remained stationary over the hot-spot, each eruption of lava built up on the previous one and Olympus Mons just kept growing.

The other volcanoes in the Tharsis Bulge region of Mars are also huge compared to those on Earth, typically standing 18km (11 miles) above the surrounding surface. Similarly located in Tharsis is a vast volcanic structure called Alba Patera. It is not a shield volcano, but rather a slightly raised section that is 1500km (932 miles) in diameter. With its large caldera and surrounding circular pattern, Alba Patera is a wonder that seems to be unique to Mars. The whole Tharsis region is located on a swelling

Looking down on Mars, the sharp cameras of the European Space Agency's Mars Express orbiting spacecraft captured this stunning view of the summit of the largest volcano in the solar system. Known as Olympus Mons, this shield volcano covers an area greater than the entire Hawaiian volcano chain on Earth and is three times taller than Mount Everest. The overhead view shown here is of the caldera, which is about 80km (50 miles) across, with pits that are 3km (1.9 miles) deep.

of the crust. This upward push of the mantle below is a plausible explanation for the growth of giant volcanic features in this region of the planet.

Lunar maria

The Earth's Moon doesn't have any large volcanoes or mountains. However, much of the surface that faces the Earth is covered in ancient lava flows. Like Mars, volcanism on the Moon is very old, resulting in plains that formed more than three billion years ago. These plains are seen today as the dark patches so familiar to us when we gaze at the Moon in the night sky. Ancient astronomers mistakenly thought these patches were seas of water and they were named "maria", a latin word meaning "seas". Today we know that these dark, roughly circular patches are not oceans but vast flat areas formed by lava flowing during the early history of the Moon. The maria solidified more than three billion years ago and the Moon has been essentially volcanically dormant ever since.

There are 19 major maria on the Moon, each a low-lying basin, with sizes ranging from 200km (124 miles) to 1200km (745 miles) across. They record the flows of numerous overlapping eruptions and cover in total about 17 percent of the surface of the Moon. The maria are, however, predominantly found on the near side of the Moon, and they cover barely two percent of the far side. This dearth of volcanic plains on the lunar face hidden from Earth's view is thought to be because the crust is higher and thicker there than on the Earth-facing side. This means that it is much more difficult for the magma to reach the surface (and fill basins) on the far side of the Moon.

Observations from Earth and the historic Apollo missions have revealed a few other volcanic features on the Moon aside from the maria. There are, for example, vents in the form of domes and cones, each a few miles across and hundreds of feet high. You could consider them as scaled-down equivalents of shield volcanoes on Earth.

Other interesting volcanic landforms on the Moon are narrow valleys usually seen in the maria, known as sinuous rilles. It is thought that they are channels carved out by tubes of lava. They can stretch for hundreds of miles, with widths from a few metres/yards to 3km (2 miles). The largest of these valleys, named Hadley Rille, was visited by Apollo 15 astronauts. In December 1972, Apollo astronauts Harrison Schmitt and Eugene Cernan made the important discovery of orange-coloured soil on the Moon. The astronauts scooped up the soil and brought samples back to Earth. Laboratory analyses subsequently revealed that the orange tints were caused by tiny glass spheres that were rich in the element titanium. After some debate, scientists now generally believe that the titanium-rich glass has a volcanic origin. It seems that the soil had been buried in lava flows until about 30 million years ago, when a rocky projectile from space struck the Moon and blasted out a bowl-shaped crater (called Shorty Crater). The discovery is of significance because lava deposits tell us about the interior history of a planet or moon. It is likely that these titanium droplets were scattered across the lunar surface by fountains of lava. It may even be that the ancient Moon was covered by oceans of magma up to 100km (62 miles) thick, which solidified to include a titanium-rich layer.

The main difference between the Earth and the Moon is of course that volcanic activity is still current on our planet, and many of Earth's volcanoes are younger than 100,000 years. Another important contrast is that while volcanoes on Earth commonly erupt in mountain chains, such as the Hawaiian Islands, the lunar events occurred in the bases of very large and old craters. Finally, water is an important ingredient of the lavas of Earth, which is thought to help increase the violence of the volcanic eruptions. Because the Moon is considerably drier than Earth, the lava there would have flowed more placidly and smoothly onto the surface.

This unusual false-colour mosaic of Earth's Moon was assembled from 53 separate images taken by the Galileo spacecraft en route to Jupiter. The colours are used to highlight dramatic differences in composition over various regions. In particular, the blue to orange shades trace the extensive, ancient volcanic lava flows. The Moon has been essentially geologically inactive for the past 2.5 billion years.

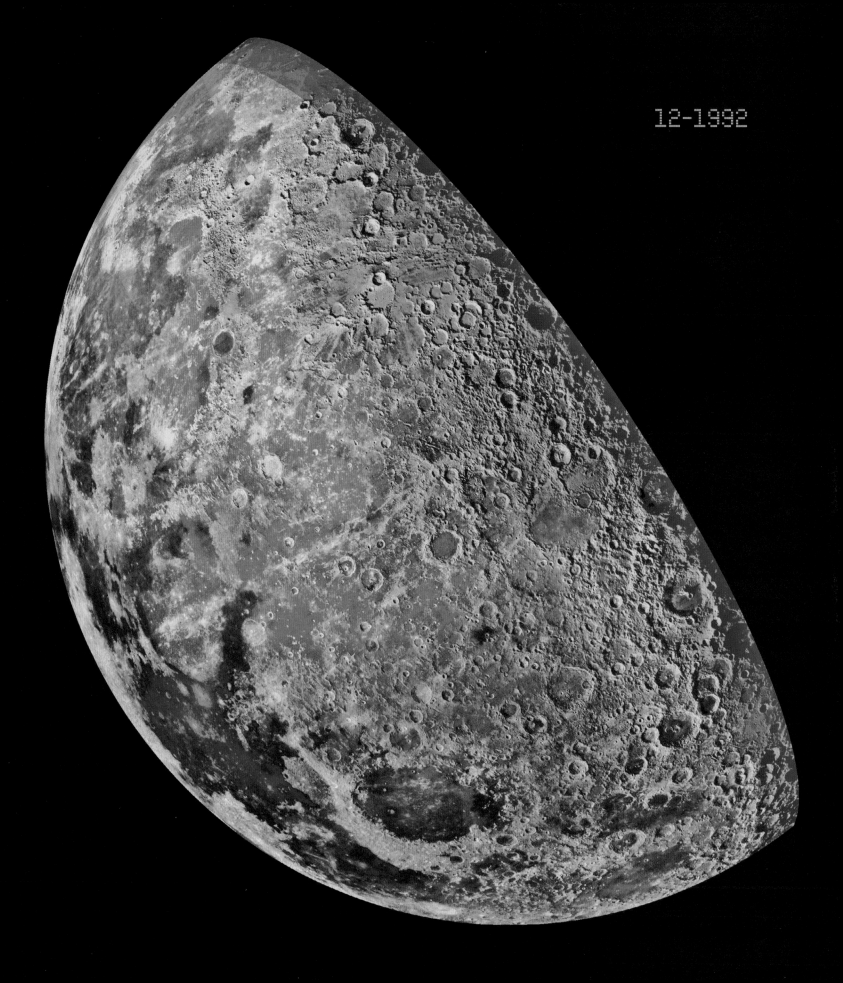

Exotic Io

One of the most awe-inspiring wonders of the solar system is Jupiter's moon called Io. This enigmatic satellite is the most volcanically active body in the solar system today. Io is one of the four large inner moons of Jupiter, discovered by Galileo in 1610. It's roughly the same size as Earth's Moon, but that's about where any comparison to our Moon comes to an end. Volcanic eruptions on Io are continuous and very common, so much so that its entire surface can be covered with 100m (330ft) of ejecta every one million years.

New space missions such as those by the Galileo spacecraft have returned striking records of active volcanoes spewing plumes of sulphur to great heights. The landscape of Io is peppered with vents, vast mountains, and great volcanic rings, giving this moon an overall collage of yellow, orange, red, and brown colours; it almost looks like a pepperoni pizza! In the late 1990s, the Galileo spacecraft observed more than 100 erupting volcanoes on Io, with some turning on and off within weeks. The

A superb, high-resolution, global image of Jupiter's volcanic moon Io. This view from NASA's Galileo spacecraft reveals a complex surface marked by fissures, lava flows, and deposits of sulphur. The unique surface is kept very young by the prolific set of volcanoes, which makes Io the most volcanically active body in the solar system.

volcanic vents are the hottest regions in the solar system away from the Sun. Volcanic temperatures in excess of 1520°C (2768°F) have been recorded – heat of this level has not been found on Earth's surface for two billion years. What is remarkable, however, is that despite the fiery volcanoes and sizzling vents, most of Io's surface is very cold indeed, and can be as low as -150°C (-238°F).

It's worth taking a close look at some of the more prominent examples. The most powerful volcano on Io is Loki, which emits more heat than all the volcanoes of Earth combined. Temperatures in excess of 560°C (1040°F) have been recorded in its vast caldera, which is constantly flooded with lava from molten material beneath this moon's crust. Another site of violent eruptions is Pele, where plumes have been ejected 600km (373 miles) above its vents. The volcanic gases spewed into the very cold air quickly freeze and condense, turning into frozen sulphur dioxide. Dramatic volcanic activity from Pele was monitored between 1997 and 1999, with lava flow temperatures of 1090°C (2000°F) recorded by the Galileo mission to Jupiter. These temperatures are comparable to those of flows seen in volcanic eruptions on Hawaii. The "old faithful" of Io's volcanoes is probably Prometheus. It has been seen to be active during a series of spacecraft observations spanning 20 years, and in images taken by the Hubble Space Telescope. This long-lived

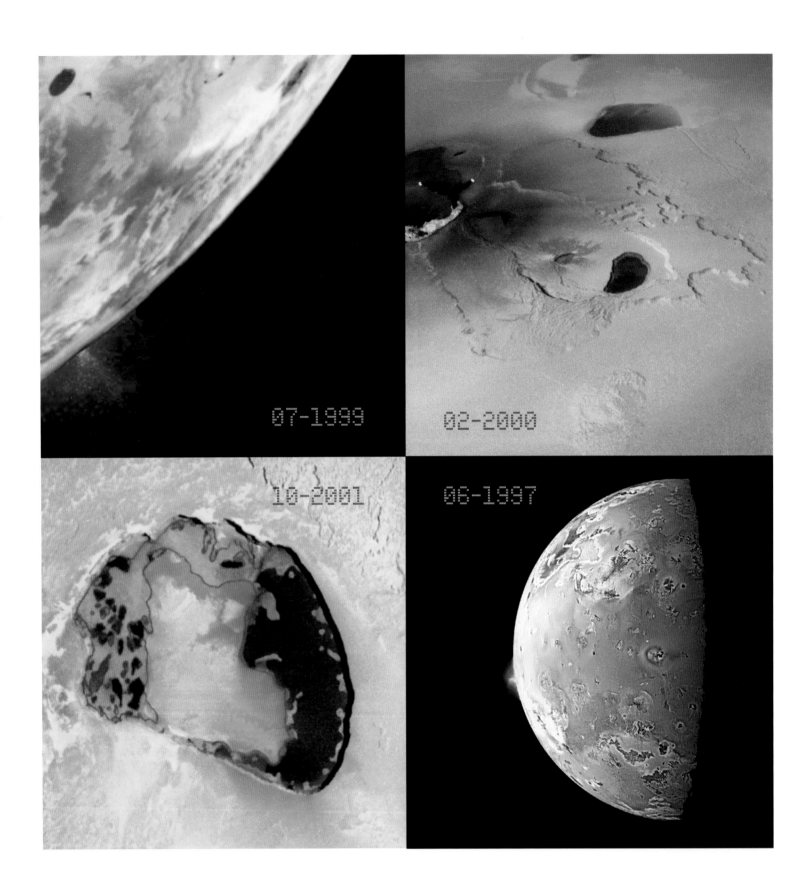

A sequence of images that capture incredible Io in action. Plumes of gas ejected over 100km (60 miles) above the surface are seen at the limb of Jupiter's active moon in the Galileo snapshots. Other prized scenes reveal newly erupted lava flowing to form bright orange-yellow rivers more than 60km (37 miles) long, together with giant caldera-like depressions that are 75km (47 miles) across. The warm lava in these volcanic "craters" is seen as black patches against the dramatic yellow surroundings that are layered in sulphur-rich compounds.

eruption has similarities with the equally persistent Kilauea volcano in Hawaii, except that Prometheus' lava flow meanders across land covered in sulphur-dioxide snow.

The lava flows on Io are mostly very fluid and travel easily along gently sloping shield volcanoes. The longest flows are probably made up of molten sulphur, which can range in colour from black to yellow, depending on its temperature. The flows can persist over remarkable distances, stretching up to 250km (155 miles) in some cases. These lava events bubble over huge, mostly circular, calderas whose dark floors provide evidence that hot lava is present. Somewhat more unique to Io are features called petra, which are abrupt fractures in the crust. The walls of these fractures can be as steep as those of the Grand Canyon on Earth, with a height of nearly 3km (2 miles).

You would expect a small moon such as Io, some 780 million km (485 million miles) from the Sun, to be a frigid and dead body, and not one that is ravaged by sizzling volcanoes. So how is the heat generated to power them? Io is too small to draw sufficient heat from the decay of radioactive elements, as is the case for our own planet. Instead Io's internal energy comes from the gravitational action of its vast planet Jupiter. As Io orbits very close to the planet, it experiences a phenomenal gravitational force. The force from the planet is combined with additional tugs from Jupiter's other moons – Europa, Callisto, and Ganymede. The effect of all this is that while one side of Io is pulled toward Jupiter, the opposite side is pulled by the

gravity of the other moons. The result is a squeezing and stretching of its interior, known as a tidal effect. Enormous stresses are created, and cause the upper layers of Io to rise and fall by 100m (330ft). The friction from this movement creates tremendous amounts of heat in Io's interior – sufficient to melt material below its crust and force it through weak points and fractures in the surface.

Io is a unique world in the solar system. Much of its forbidding behaviour remains a mystery. Scientists are keen to understand the cause of the fierce temperatures of its volcanic vents and the chemical make-up of the dynamic plumes. Detailed and sharper images need to be combined with accurate measurements to further understand Io's lava flows and the balance of sulphur and silicates in their composition. The shapes formed by the lava can teach us a great deal about its nature, but that ideally requires images that can resolve features down to several yards on Io's surface. Only flybys from spacecraft such as Galileo can provide such detailed information. The data will ultimately permit more rigorous comparisons with Earth, thus advancing our understanding not only of the prolific activity of Io, but also of our own currently dynamic planet.

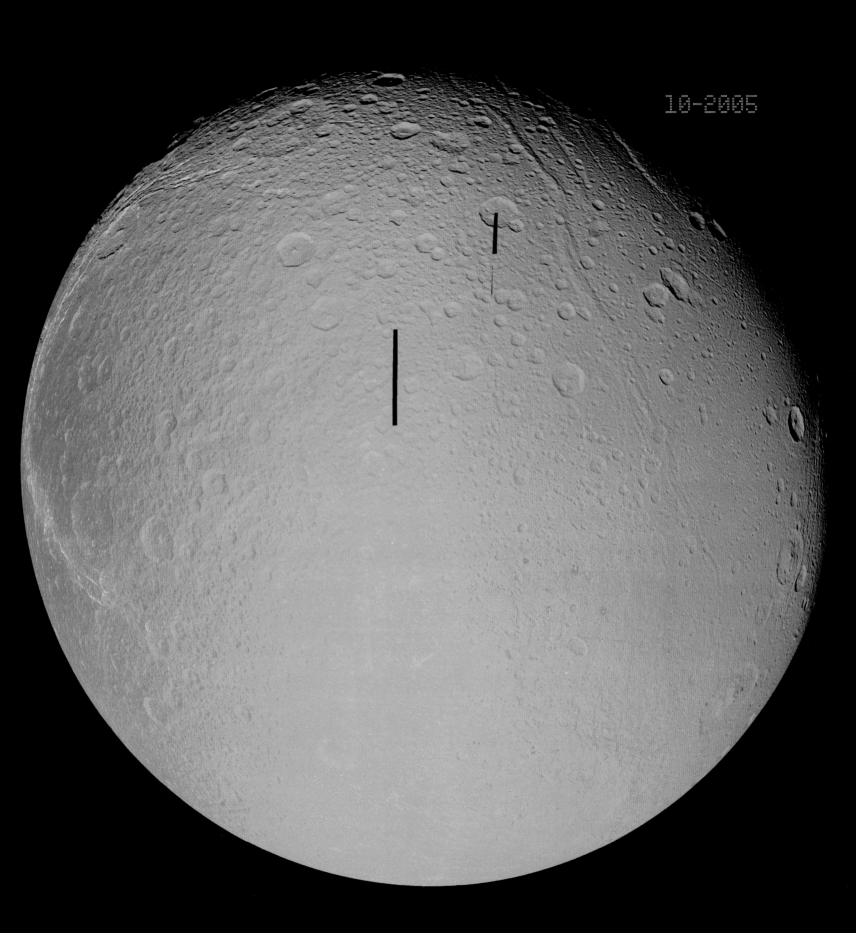

11-2005

10-2005

Ice-cold Eruptions

The terrestrial notion of volcanoes usually conjures up images of fire, violence, destruction, and immense heat. Our exploration of the solar system is, however, continually stretching our perspectives, and this includes the discovery of bizarre forms of volcanism. Among the strangest examples are ice volcanoes, where the eruptions are not of lava, but of frozen materials such as ice, methane, or liquid nitrogen. Known as cryovolcanism, this remarkable process can be witnessed on some very exotic moons of giant planets in the solar system.

One of the first examples of cryovolcanism discovered in our solar system was on Neptune's largest moon, Triton. Gravitationally captured by Neptune early in the planet's history, Triton is a Pluto-like icy body with surface temperatures as low as -230°C (-391°F). Unusually, this large moon orbits Neptune in

the opposite direction to the planet's rotation. It also spins on an axis that is highly titled compared to Neptune's axis. The consequence of this unique setup is that Triton experiences seasons, with significant temperature changes as its poles take turns to face the Sun for extended periods of time. The seasonal changes can be enough to raise the temperature beneath Triton's thick nitrogen ice surface. The deep layers absorb light and trap heat, causing the nitrogen to melt. A 10°C (18°F) rise in temperature can, for example, cause the sub-surface nitrogen to expand a hundred fold. The expansion builds up enormous pressures until pockets of gas explode, releasing nitrogen geysers that shoot 8km (5 miles) above the surface. The plumes rapidly freeze again surrounding the region in nitrogen snow. Much like lava flows on Venus or the Earth, large areas of Triton have been re-surfaced by the eruption of these molten ices. This remarkable moon also

These views from Cassini show two of Saturn's ice-encrusted moons. On the left, fountains are seen erupting from the ice volcanoes of Enceladus. Water vapour from warm ice rises 400km (250 miles) above the surface. Dione, right, is seen against Saturn's rings. Dione is 11,000km (6835 miles) across, with a dense composition of water ice and rock.

Previous page
Left: an infrared view from the Hubble Telescope showing a rare alignment of Jupiter and two of its moons, with three shadows. Io is the central white dot (its shadow is to its left). The blue dot, right, is Ganymede (its shadow is far left. Top right of the image is Callisto's shadow.
Right: Cassini infrared and ultraviolet images have been processed for this false-colour mosaic of Saturn's moon Dione.

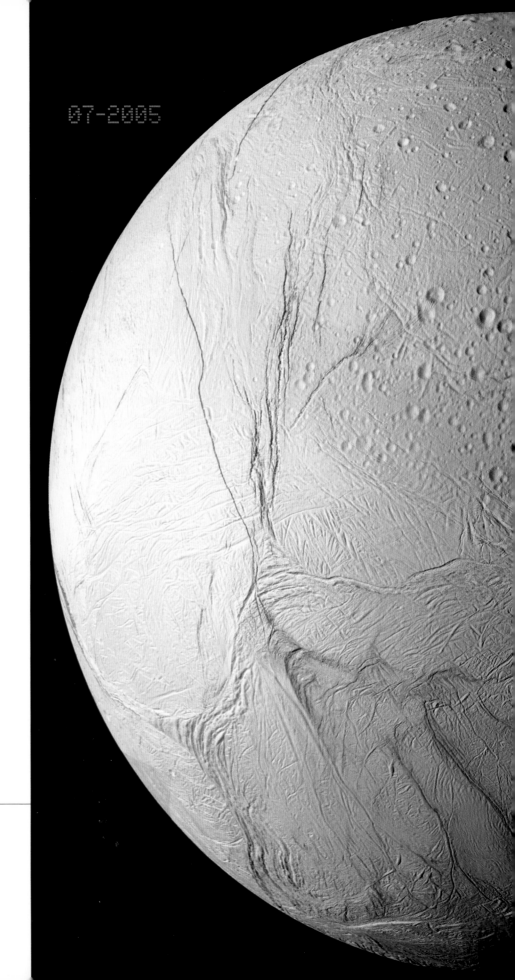

has other features traditionally associated with volcanic activity, such as calderas, built up layers of (icy) crust, and deep depressions in the land that are separated by ridges.

Another fascinating location for ice volcanoes is the solar system's largest moon, Ganymede. One of Jupiter's major moons, Ganymede is larger than the planet Mercury. Most of the surface of this moon is made up of icy plains. Detailed images from the Galileo spacecraft also reveal a rugged surface, with mountains and sheets of ice. There is evidence of water eruptions on the surface of Ganymede. Particularly interesting are extensive troughs in the surface that descend for a mile or more, somewhat like the Great Rift Valley on Earth. The walls of Ganymede's valleys are very smooth, however, and provide evidence that fairly fluid, warm ice once flowed there. Flooded by slush from icy volcanoes, Ganymede also exhibits depressions that resemble calderas, together with fractured or wrinkled features caused by the crust being pulled apart or pushed together. All these imperfections could be potential sites for sub-surface water to seep out, travel, and then freeze to create the vast plains of ice that now appear to cover the rough terrain and ancient craters.

In late 2004 and early 2005, the Cassini spacecraft flew by Saturn's enigmatic moon Titan and revealed evidence of more exotic volcanism. Using special infrared cameras to penetrate the moon's thick atmosphere, a 30km (19 mile) wide circular feature was revealed. The structure had two wings extending from its westward side, and the whole feature has been interpreted as the dome of an ice volcano. It resembles volcanoes on Venus and Earth, where overlapping layers are created from a series of lava flows. There is even a dark central patch on Titan's volcano that may be a caldera. The material expelled is likely to be methane ice, plus other chemicals such as ammonia. As with Triton and Ganymede, an internal (sub-surface) heat source is thought to drive the volcano. The cryoeruption on Titan might even have resulted in extensive methane rains that carved out channels along the slopes of the

Detailed features of Saturn's moon, Enceladus, are revealed in this high-resolution image acquired by Cassini. The terrain of this ice world is a mixture of fractures and craters. The surface is as bright as fresh snow and its ice volcanoes are thought to rise like water geysers from the surface gashes, dubbed "tiger stripes". The icy eruptions create a weak atmosphere around this moon and provide particles for Saturn's outer ring.

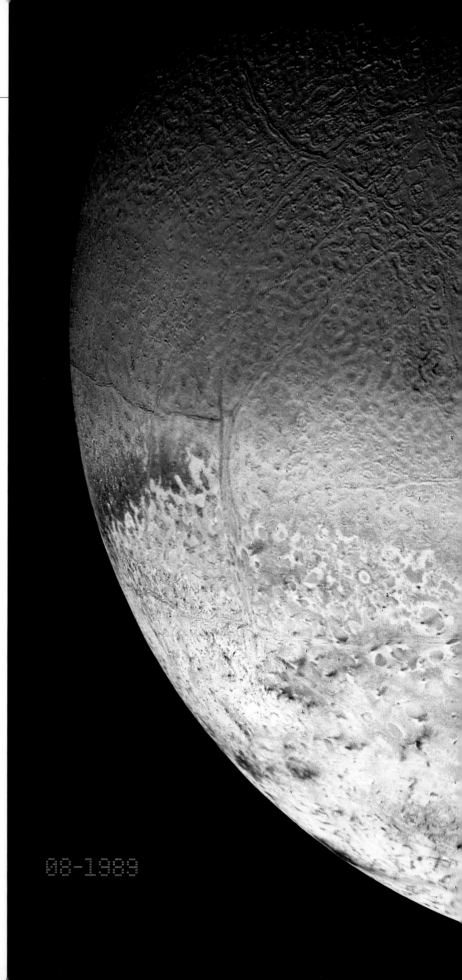

Voyager 2's encounter with Neptune's largest moon, Triton, revealed a complex world with a plethora of features, including small cryovolcanoes that spew out a mixture of nitrogen frost, ice, and organic compounds. The plumes are thought to be a widespread occurrence on Triton, perhaps carrying with them darker material from the crust. The moon is 2705km (1680 miles) across, and has an average surface temperature of -230°C (-391°F), making it one of the coldest bodies in the solar system.

volcanic dome. Methane is a very abundant gas in Titan's dense atmosphere. However, because of the moon's low gravity, and bombardment with ultraviolet rays from the Sun, the molecules are expected to break up and disappear. So something must be continually producing and topping up the methane. The source of this extra methane was a mystery, but now it seems that it may be produced by the plumes released from icy volcanoes.

The Cassini spacecraft has also visited another of Saturn's remarkable moons, called Enceladus. Despite being barely 500km (310 miles) across, this tiny moon has a fascinating surface that includes craters, very smooth low-lying plains, long stretched ridges, and narrow valleys. What is remarkable is that Enceladus appears to have a very bright and relatively fresh icy surface. Water frost has continually covered the orb to make it highly reflective of sunlight; in fact, it is the most reflective of any moon in the solar system. The suspicion is that Enceladus is being re-surfaced by water-ice eruptions driven by internal heat. The energy is likely to come from gravitational stresses between the tiny moon, Saturn, and Saturn's other satellites. In 2005, the Cassini spacecraft also revealed that this minuscule moon had an atmosphere of volatile hydrocarbons and water ice. The gravity of Enceladus is simply not enough for it to hold on to an atmosphere for long. There has to be a continuous source from below, therefore, to pump up the atmosphere, which is now thought to be icy eruptions from volcanoes and geysers. We will see later in this book that all these fascinating sites of cryovolcanism in the solar system also offer tantalizing possibilities for the presence of primitive sub-surface life forms (see pages 128–31).

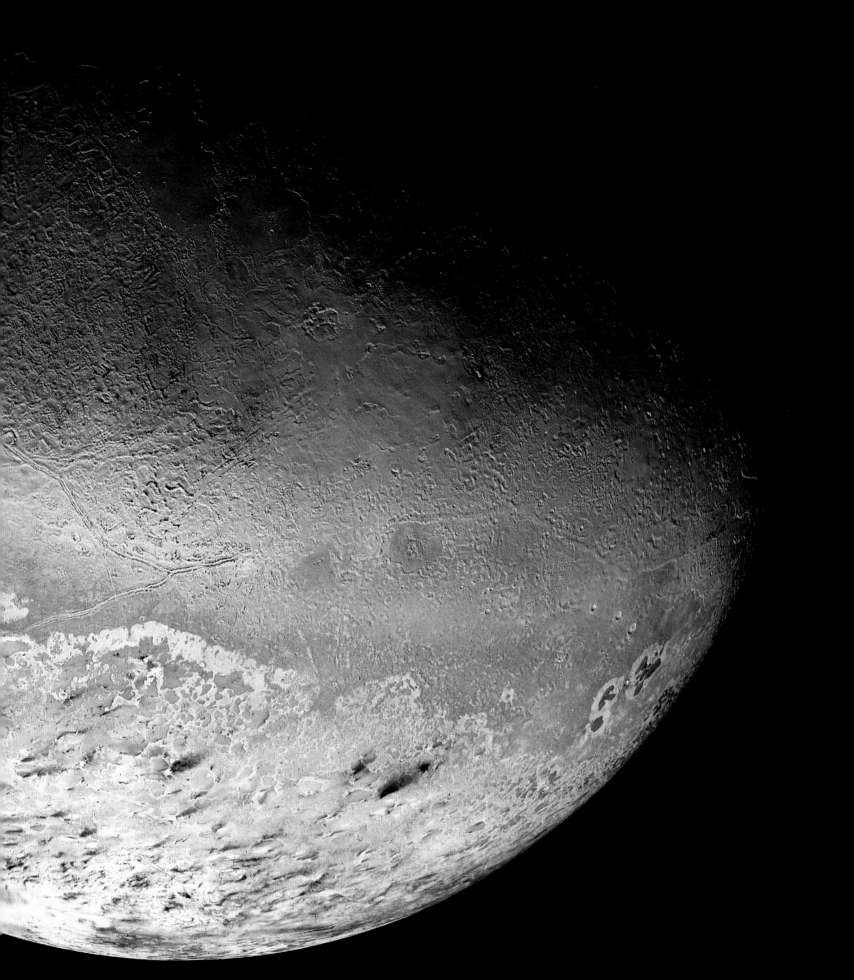

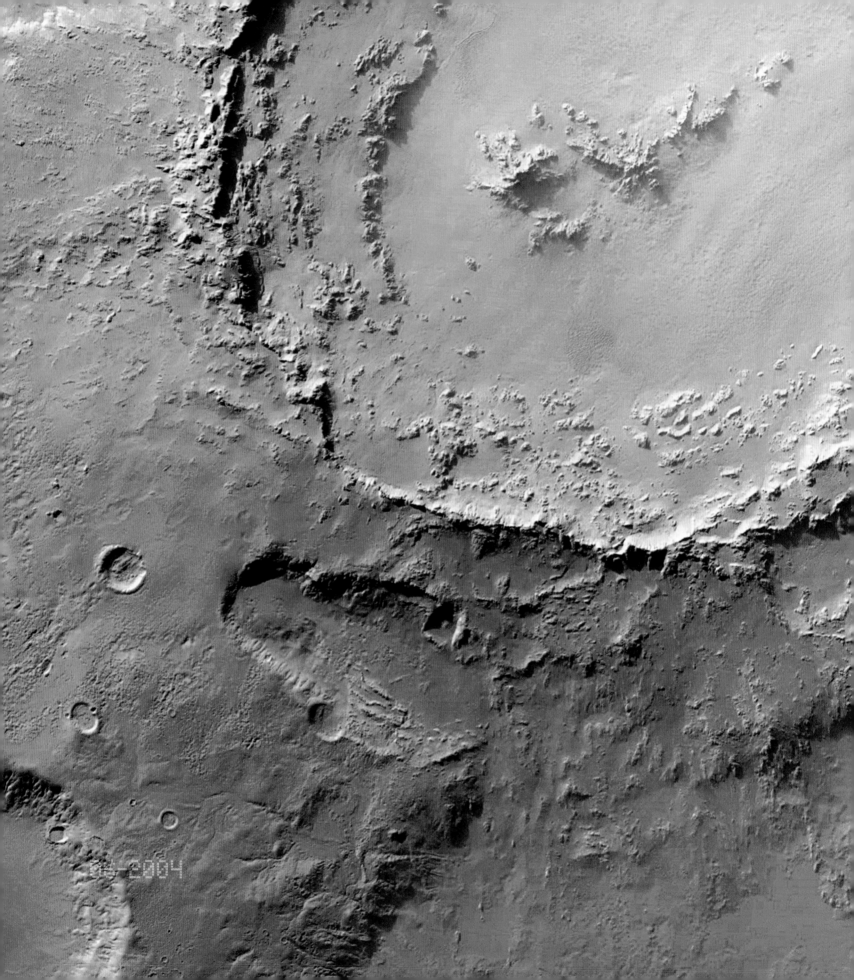

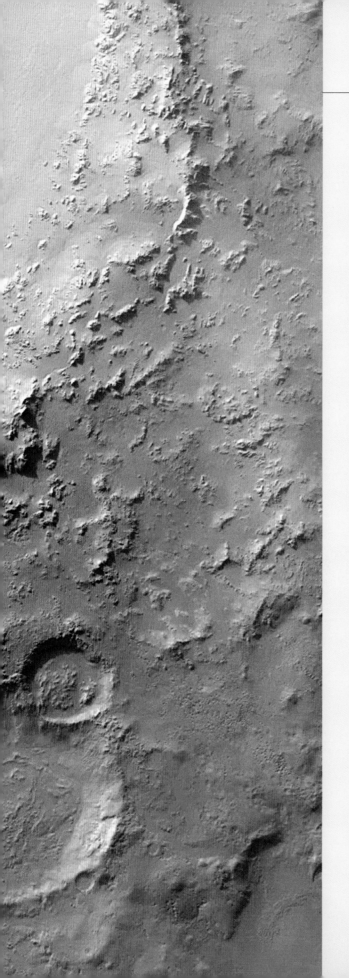

The high definition camera on board the European Space Agency's Mars Express spacecraft delivered this detailed image of the Hale impact crater on the southern hemisphere of Mars. The crater is 136km (85 miles) wide and displays classic crater features, such as terraced walls and mountainous central peaks. Surface features down to a resolution of 40m (130ft) per pixel are revealed in this crisp image.

Impacts and Craters

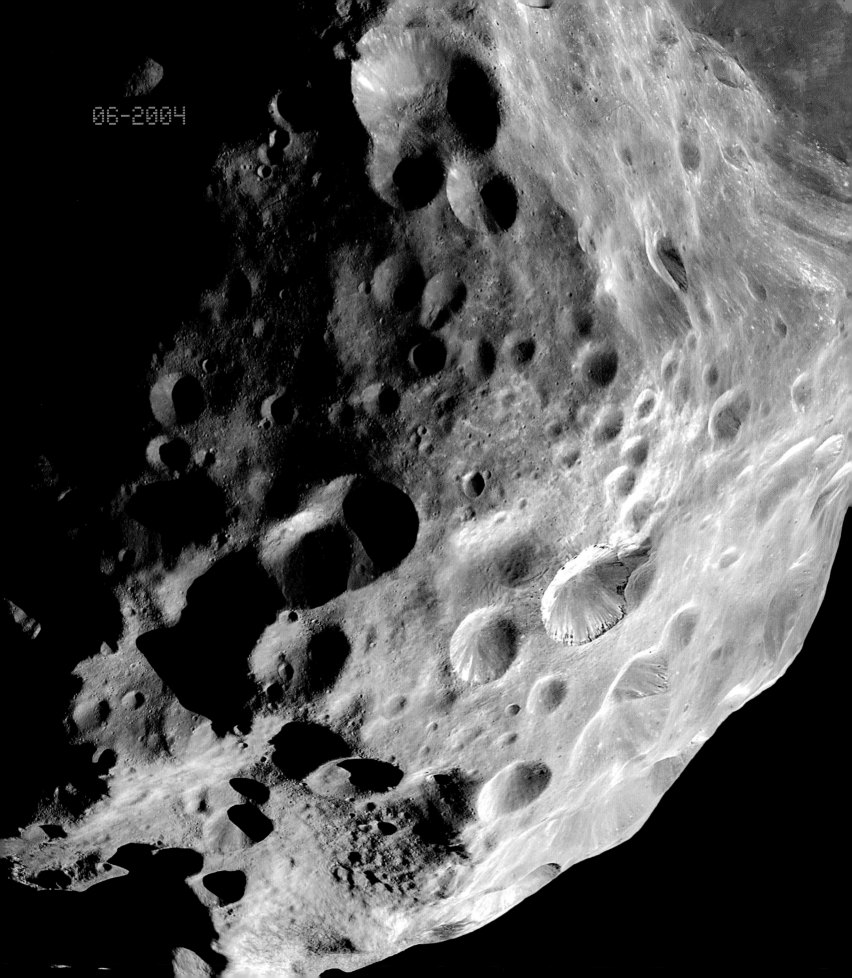

All the inner planets and the numerous moons of the solar system have been heavily bombarded by objects such as asteroids and comets throughout history. The scars left by this violent battering are still seen today as impact craters. This catastrophic process has played a dominant role in shaping the geological histories of the planets. The projectiles from space crash at high speeds, creating massive disturbances in each planet's crust and exposing huge volumes of the deepest materials. On Earth, impact events also play a significant role in the evolution of life, having led to at least one major biological extinction. This chapter looks at the diverse and spectacular results of cosmic collisions evident on the Moon and Mercury, together with the more masked craters of Mars, Venus, and Earth. Though far less frequent, catastrophic impacts are still occurring in the solar system today. For example, the collision in July 1994 between a comet and the giant gas planet Jupiter, and the suicide comets that plunge into the Sun. These remarkable events provide reminders that the possibility of collision between extra-terrestrial objects and our planet still poses a real, though very rare, threat to humanity.

Saturn's intriguing moon Phoebe can be seen in this stunning image from the Cassini spacecraft. The view is of a remarkably irregular terrain, scarred by numerous craters. The 220km (137 mile) wide moon exhibits extensive variations in brightness, thought to be due to bright ice on some of the crater floors and walls. Phoebe may once have been an icy, comet-like body residing well beyond Neptune that became trapped in Saturn's gravity during a past foray into the solar system.

Making Craters

Around four billion years ago when our solar system was still forming, it was a very dangerous place indeed. It was a cosmic shooting gallery, with millions of rocky objects speeding in different directions. During the first half-billion years, the surfaces of the newly formed planets in the inner solar system were continually pulverized by crater-forming boulders of various sizes. The results of these collisions are still visible as vast circular basins and giant holes.

The landscapes of Mercury, Earth's Moon, and. to some degree. Mars, have preserved the record of this bombardment, primarily because they have remained unchanged for millions of years. The relatively larger planets Earth and Venus, with their greater gravitational pulls, would have attracted an even heavier bombardment in their early histories. However, impact craters are not so obvious on Earth because it is so geologically active.

The effects of mountain building, volcanism, and erosion have transformed or buried the impact sites, and there are only around 160 cosmic craters currently identified on our planet. Comparisons with Mercury and the Moon suggest that the Earth would have been targeted to produce more than three million impact craters, ranging in size from 1km (0.6 miles) to more than 1000km (620 miles) across. Given that the most prolific rate of bombardment persisted until 3800 million years ago, it is unlikely that life could have taken hold on Earth any earlier.

There is no doubt that large impact phenomena played a major role in the early formation of the solar system. The orbits, compositions, and atmospheres of newly formed planets may have been affected by these events – indeed, the currently favoured theory for the formation of our Moon is that it is the result of a massive collision. More than four billion years ago, a

This freshly excavated crater on Mars was not formed by a rocky body from space, but was created instead by the crash of the heat shield of the craft that carried the Mars Exploration Rover. The heat shield was jettisoned after entering the Martian atmosphere and struck the ground to form a "man-made" crater about 2.8m (9.2 feet) wide. The impression in the terrain (on the right) is barely 10cm (4in) deep and reveals reddish sub-surface material. The main part of the remnant heat shield is seen on the left of the image.

Mars-sized object struck the proto-Earth in a glancing impact that blasted debris into orbit around our planet. This material subsequently gathered and clumped under gravity to form the Moon. Similarly powerful impacts may have altered the primordial atmospheres of planets, ripped away the mantle material of Mercury, or forced Uranus' axis of rotation to be tilted more than 90 degrees from the axes of the other planets orbiting the Sun.

Surface-scarring craters are, however, the most common geological structures resulting from impacts. The comparison and study of craters throughout the solar system provides a valuable and detailed record of events. Of particular value are the well- preserved craters on planets where erosion and volcanic activity have not been extensive; the surface of the Moon in particular is marked with numerous examples that provide a basis for understanding impact events. The crashes can

also expose the deepest layers of a planet's crust, and even on Earth impact sites have been rich in commodities such as diamonds, uranium, oil, and gas. Incoming comets or asteroids may even deliver organic compounds from space, though these are probably destroyed in the harsh conditions of planets such as Mercury, Mars, and Venus.

Extensive observations of lunar craters, and examination of key sites on Earth, enable us to build up a detailed account of impacts in action. The violent nature of an impact event is principally due to the high speed with which an asteroid or comet strikes the terrestrial planet. Typically, the impactors arrive on the Earth with cosmic velocities of 10–70km per second (up to 45 miles per second). These terrific speeds make space projectiles considerably more explosive than TNT. An object barely a few yards across would carry the moving (kinetic) energy equivalent

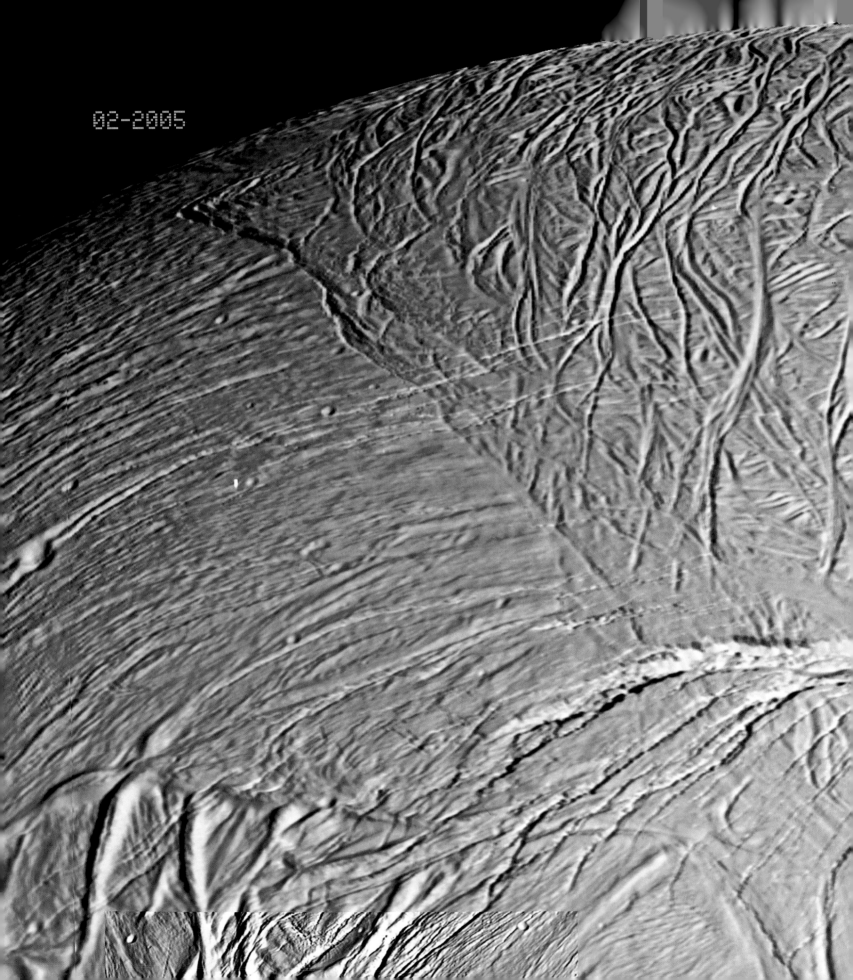

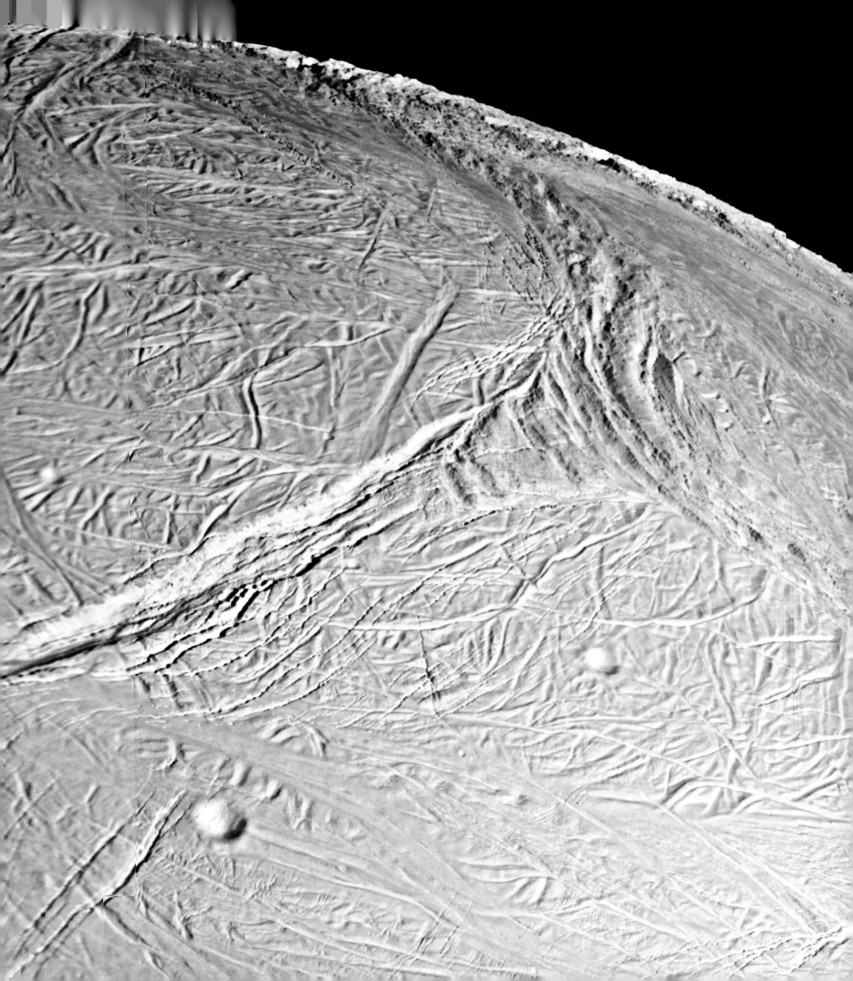

to that of an atomic bomb. The impact of a kilometre-sized body would release more energy than all of the Earth's geological processes combined have yielded in hundreds of years. The key difference, however, is that the immense energy from the impact is released in a matter of seconds as opposed to the longer timescales of volcanoes and earthquakes.

The effect on the planet's surface is instantaneous. As soon as the impactor strikes the surface of the planet, the kinetic energy is channelled into a very strong shock wave that travels rapidly through the rocky material. The rock is destroyed, melted, or vaporized. The explosion excavates a previously flat layer of rock to form a bowl-shaped crater that is surrounded by ejected materials. The compression and excavation are over in seconds, making impact cratering a prolific geologic process. Most often the shock waves created are also powerful enough to melt the incoming projectile. The size, shape, and complexity of the resulting crater can depend on the mass, speed, and angle of arrival of the projectile.

Broadly, impact craters are usually divided into two categories; simple craters and complex ones. Simple craters are smaller, with smooth walls and a depth of about 20 percent of the diameter. Complex craters are formed in more ferocious events, where steep excavated walls collapse inward to form single or multiple peaks in the middle. Further modifications can occur as the deep crater floor rebounds from the shock wave of the impact. These complex craters can be as much as 100km (62 miles) across, but they are relatively shallow compared with their diameters.

A great cataclysm

Though planetary bodies have been pummelled with rocky and icy projectiles throughout their histories, it appears that there was one particularly spectacular assault on Earth, its Moon, and the inner planets. This took place about 3.9 billion years ago, and lasted 20–200 million years. Almost 80 percent of the Moon's crust was resurfaced, and more than 1500 giant craters, each at least 100km (62 miles) across, were created. Evidence for this mega-pounding comes from lunar samples brought back by the Apollo astronauts, which revealed that the Moon's surface was subjected to extreme heat 3.9 billion years ago.

The Earth would not have been spared this beating, with possibly tens of thousands of craters being formed. This great cataclysm would have had a critical effect on the early development of life on our planet. So where did all these impacts come from? The most recent consensus is that it was due not to comets, but rather to a flood of asteroids. The bombardment probably originated in the main asteroid belt, which is still present between the orbits of Mars and Jupiter. The pock-marked scars of craters in the older regions of the inner planets (especially Mercury and Earth's Moon) are remarkably similar in size to what would have been expected for the range of bodies in the asteroid belt. (This belt would have been in place more than four billion years ago.) Thankfully, it is highly unlikely that the asteroid belt will again unleash its ammunition. The solar system is considerably more settled today, though there are still some rogue objects that we need to watch out for (see pages 104–6).

This image from the Mars Global Surveyor shows a set of three impact craters on Mars. The rare structure probably formed when a single body, such as a meteor, broke into three pieces and struck the Martian surface. The full span shown here is about 3km (1.9 miles) across. These impact craters are coincidentally on the floor of another much larger crater.

Previous page
A beautiful view of Saturn's ice-encrusted moon Enceladus. The image was taken by the Cassini spacecraft from a distance of 17,400km (10,810 miles). About 300km (186 miles) of intriguing terrain is shown here, with features that include tiny impact craters, long fractures, and folds in the surface. A large part of Enceladus' surface is covered in craters, many of which are deformed or softened because they were carved in water ice. The older and more cratered regions of this moon contrast with areas that show no craters at all, likely due to resurfacing from water ice in geologically recent times.

Overleaf
Parked on the rim of Erebus crater on Mars, the Exploration Rover Opportunity assembled this outstanding panoramic view, based on 635 separate images. The crater is about 300m (984ft) across. A very rich variety of features is apparent around this relatively shallow crater, including small cobbles, larger rocks, and ripples formed by the action of winds. Erebus is a rather old and eroded crater that is today difficult to discern from the surrounding land.

11-2005

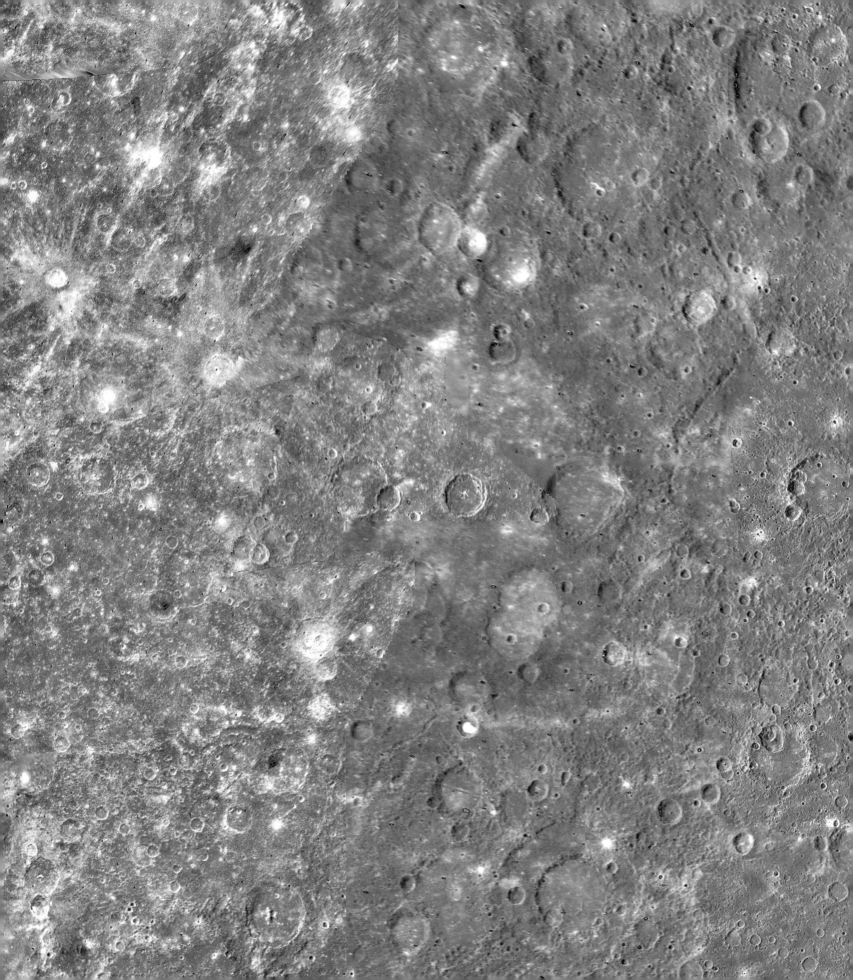

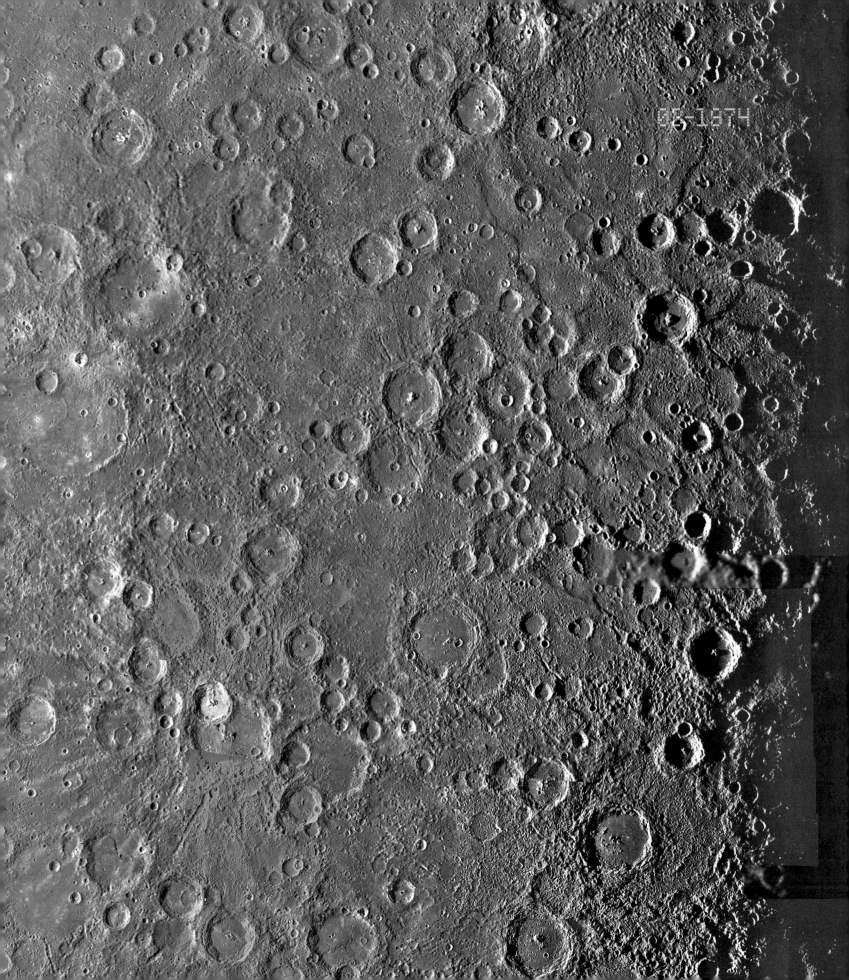

A Crater Tour

Telescope and spacecraft observations have revealed impact craters on every solid planetary surface in the solar system, including Mercury, Venus, Earth and its Moon, Mars, and the satellites of the giant gas planets. Even the asteroids themselves are often scarred by impacts. The amount of cratering is normally an indicator of the age of the surface. The oldest surfaces in the solar system are marked by the most craters. In contrast, the geologically active planets have been more recently resurfaced, by volcanic flows, for example, and exhibit fewer craters.

With its proximity to us, lack of erosion from wind and water, and its very minimal geological changes, Earth's Moon offers an ideal opportunity to study pristine craters. The Moon still has features dating back almost to its formation. Its surface is marked with millions of craters that range in size from thousands of miles to inch-sized (and smaller) micro-craters. The greatest craters have mostly been flooded by lava to form the maria discussed in the previous chapter. Only the rims of some of these basins can be seen now as cliffs and scarps.

Lunar craters can be a dazzling sight through binoculars or a small telescope, particularly if they are close to the day-night boundary of a crescent Moon. Two superb examples are the relatively young Copernicus and Tycho craters, both of which are set amid an extensive display of rays made by the light-coloured debris blasted out during crater forming impacts. Copernicus is a prominent feature, northwest on the Moon's Earth-facing hemisphere, and is 93km (58 miles) across. It probably formed from an impact about 800 million years ago, with debris spread across almost 800km (500 miles). The floor of the crater has peaks over 1km (0.6 miles), with the inner walls broken into terraces. The crater Tycho is the youngest large impact crater on the Moon's nearside. Ejecta from this 85km (53 mile) wide crater was scattered over vast regions. The Apollo 17 mission in December 1972 landed on one of the rays formed by this excavated material, almost 2000km (1245 miles) from the crater itself. Laboratory analysis of samples from the site suggests that the Tycho crater was formed almost 100 million years ago.

During the 1990s the Clementine and Lunar Prospector missions orbited the Moon and established that the Aitken Basin on the farside is the largest known impact crater in the solar system. A comet-sized body created a depression some 2500km (1550 miles) wide. Remarkably this crater is a 13km (8 mile) deep gouge that may have penetrated the lunar crust and exposed the Moon's mantle. For this reason the Aitken Basin represents an excellent site for future exploration of the Moon.

Despite its entirely different environment at the edge of the Sun, the planet Mercury has a remarkably similar surface to that of Earth's Moon. Lacking an atmosphere, Mercury still hosts a vast number of craters. The largest of these is the 1300km (800 mile) wide Caloris Basin, thought to be the result of an asteroid crash. The shock wave from this impact affected the terrain on the directly opposite side of the planet. The basin gets extremely hot as it directly faces the Sun during Mercury's closest approach to the Sun; hence the name Caloris, which means "heat" in Latin.

Backyard impacts

Although the scars of the ferocious bombardment of the Earth are largely buried today by our planet's continually changing geology, there are still a few fascinating examples of terrestrial craters. The majority of these are located in regions that have been spared extensive resurfacing from volcanoes, earthquakes, and land infilling. The more than 150 impact craters identified on Earth today range from mile-sized bowls to more complex structures more than 200km (125 miles) across. The first impact crater to be recognized was the Barringer Crater in Arizona, USA. It was created about 49,000 years ago when a 30m (100ft) iron meteorite struck Arizona's Colorado Plateau. Barringer is an impressive sight, with a classic circular crater and a raised rim, beyond which lie deposits of ejected material.

Some of the oldest and largest impact craters on Earth are found preserved in the icy terrains of northern Canada. Manicouagan Crater is a prime example best viewed from space. It is seen today as a 70km (43 mile) ice-covered annular moat. The diameter of the original crater was probably three times that of the current lake; the impact itself has been gradually eroded by the movement of glaciers. The crash is thought to have occurred about 212 million years ago with the destruction of a 5km (3 mile) asteroid. Another fascinating example in Quebec is the Clearwater Lakes, which are twin lakes that formed simultaneously when a pair of asteroid-sized bodies slammed into the Earth about 290 million years ago. This rare twin crater phenomenon has left lakes carved into the ice shelf that have diameters of 22km (13.7 miles) and 32km (20 miles).

These impact craters on the Earth's Moon were imaged by SMART-1, which is the first spacecraft the European Space Agency has ever sent to the Moon. The probe tested pioneering propulsion technologies, including a solar-powered ion thruster. The craters in this view are located near the northern limb of the Moon and the largest structure seen here is about 135km (85 miles) in diameter.

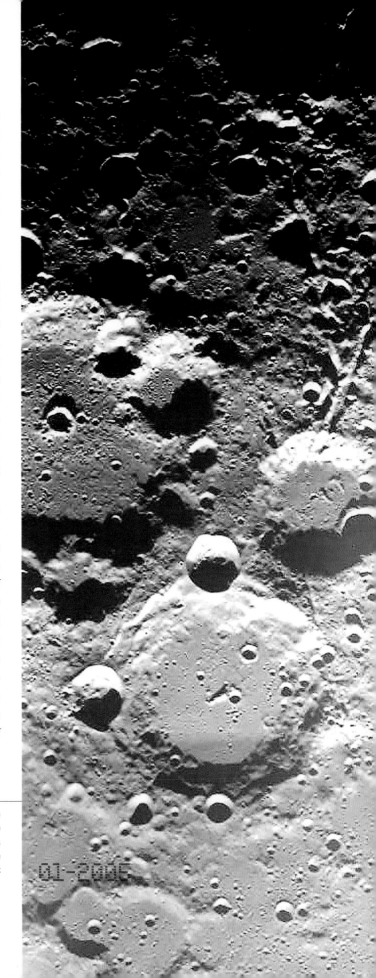

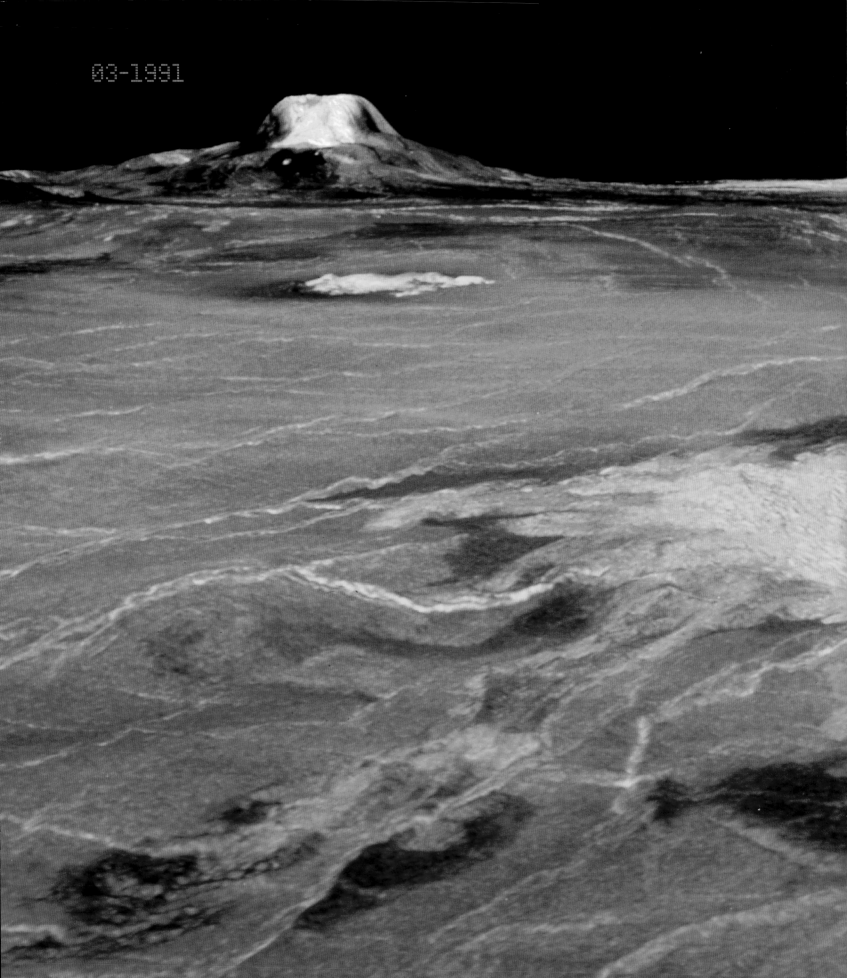

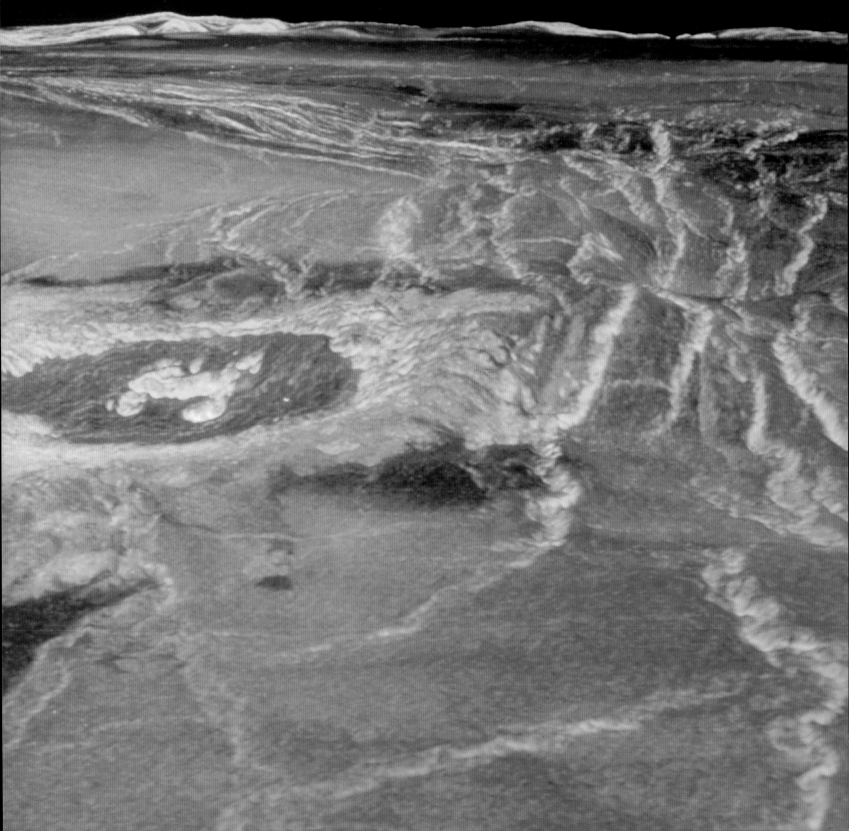

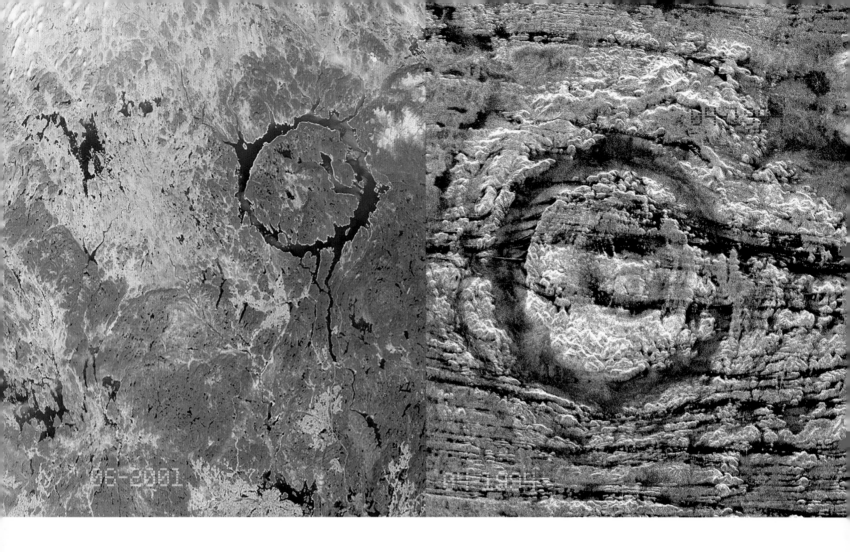

Life on Earth has, it seems, not been spared the consequences of cosmic bombardments. The most famous example is the large impact event thought to have occurred about 65 million years ago, which marks the boundary between the Cretaceous and Tertiary periods. It corresponds to one of the greatest mass extinctions in our planet's history. Up to 75 per cent of the animal species of Earth were extinguished, from the dinosaurs, to most of the plankton in the seas. The notion that this mass extinction was caused by an impacting asteroid or comet was supported by the discovery of large amounts of the chemical element iradium in rocks deposited during this period of history; only cosmic debris is very rich in iradium.

Following extensive searches, there is some consensus that the giant crater produced by the impact is the 180km (110 mile) wide structure called Chicxulub, in the Yucatan Peninsula, Mexico. It is buried deep in sediments and has been imaged with penetrating geological techniques. The asteroid or comet was about 10–20km (6–12 miles) across and debate continues as to the precise chain of events that ensued after this crash. Energy equivalent to that of 100 million megatons of TNT would have been generated, sufficient to violently excavate rock deep beneath the Earth's surface. Most of the debris would have been deposited as thick layers over North and South America. Billions of tons of sulphur and other materials were also kicked up into the atmosphere. The mooted consequence was that the dust in the atmosphere blocked the sunlight and plunged the entire globe into darkness and cooler temperatures for six months. For a good fraction of this period photosynthesis would not have been possible, drastically affecting the food chains. Other hazards to life would have included toxic vapours and sulphuric-acid rain. There is also evidence of charcoal and soot, which suggest that widespread forest fires resulted. It is likely that the Chicxulub impact set in place a set of additional related events, such as tsunami waves and fierce winds that, together with the direct effects, destroyed so many species.

Earth still bears scars from the ancient bombardment from asteroids, comets, and other rocky bodies. From left to right, this trio shows: the 70km (44 mile) Lake Manicouagan structure in Canada; the crater resulting from the crash of an asteroid or comet several hundred million years ago in Northern Chad in Africa; and the rare twin craters in Quebec, Canada, probably formed by two fragments of the same meteor.

Previous page
A view of the surface of Venus constructed from radar data from the Magellan spacecraft. There is a 49km (30 mile) wide impact crater in the centre of the image. The Gula Mons volcano is on the horizon.

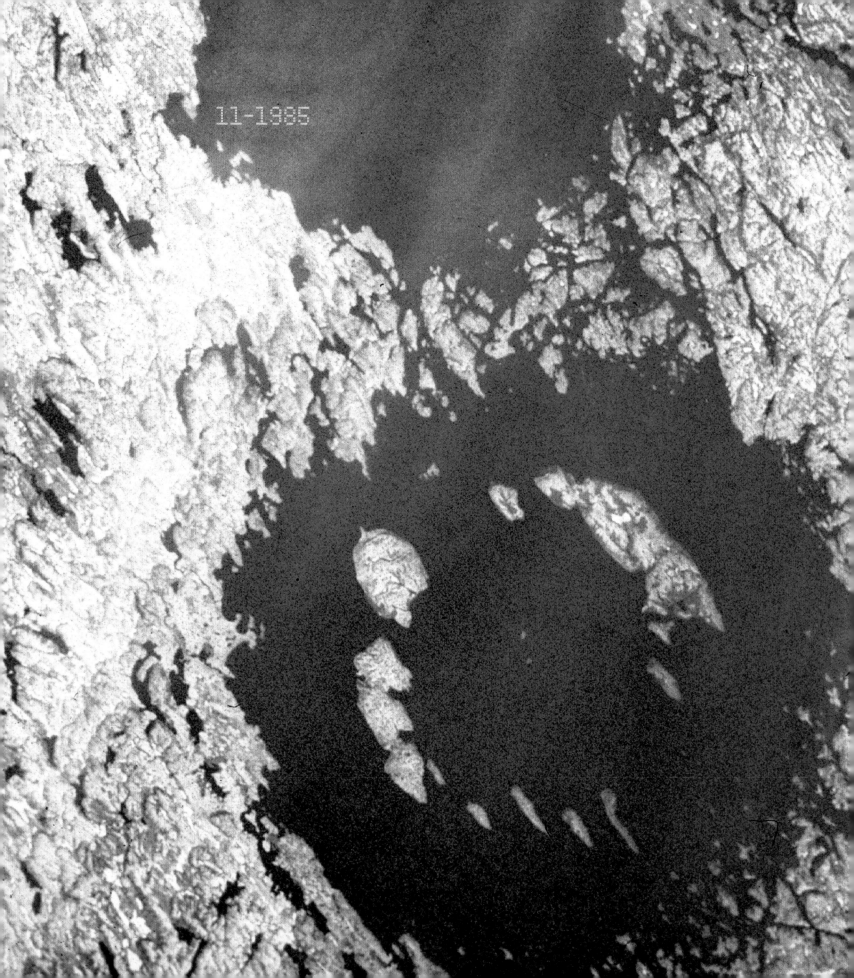

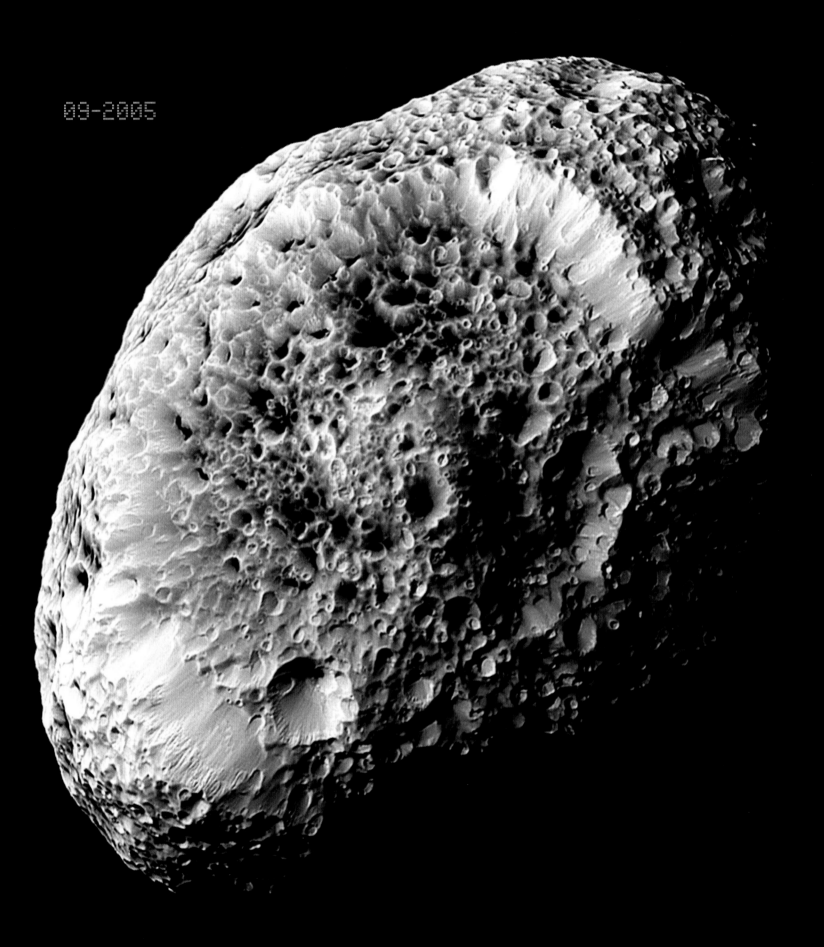
09-2005

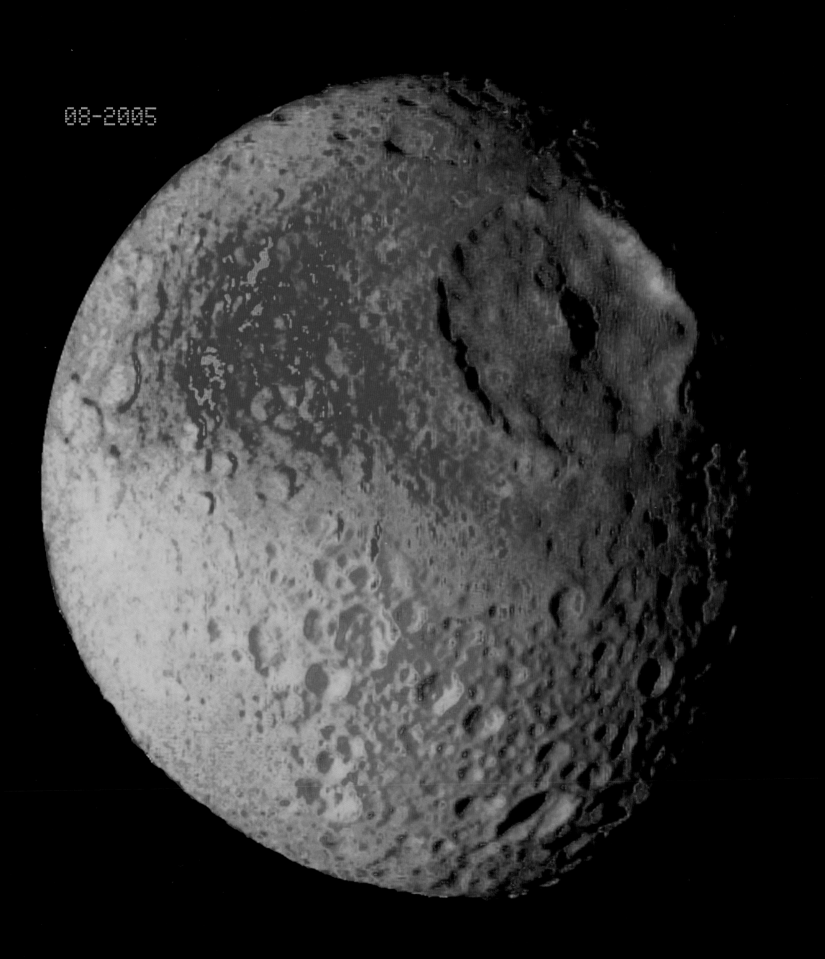
08-2005

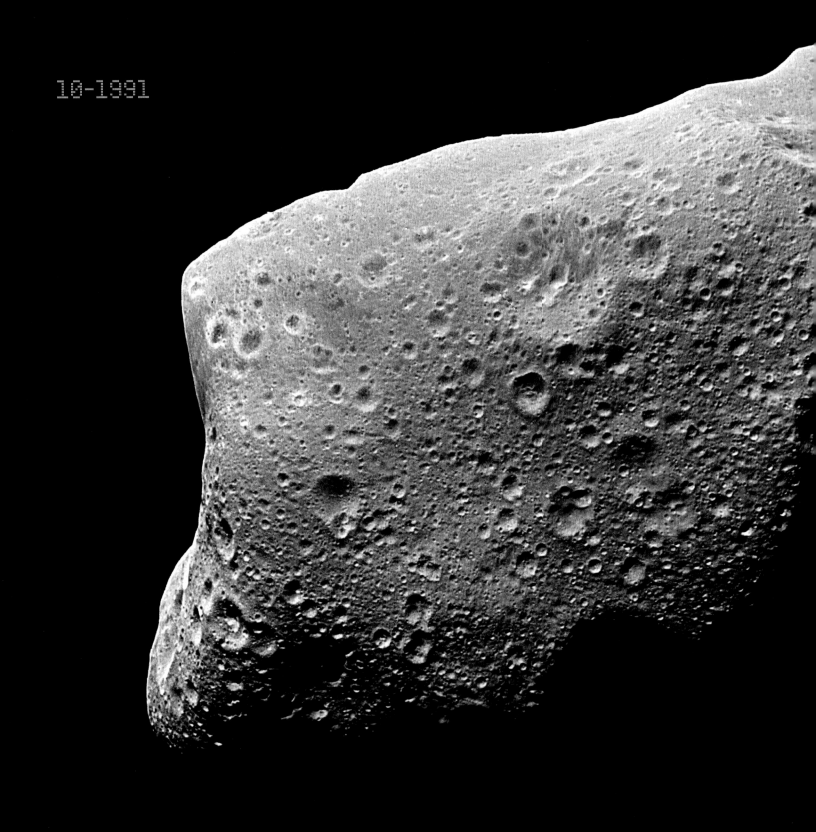
10-1991

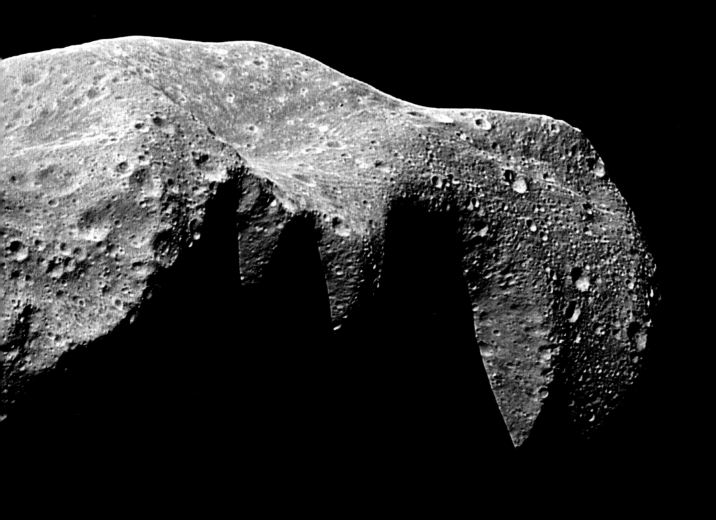
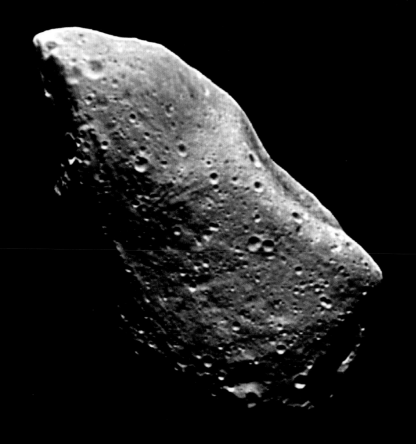

08-1993

Direct hits in the outer solar system

Aside from the plethora of craters on terrestrial bodies such as Earth, its Moon, and Mercury, a variety of well-preserved craters are also found in an exotic range of objects in the outer parts of the solar system. An interesting issue in many of these cases is that impactors were not just striking the rocky planets. In particular, the moons of Jupiter and Saturn mostly have very cold surfaces covered with ice. At these low temperatures the ice is very strong and can help preserve craters. Another important difference in the outer solar system tends to be the nature of the projectile itself. The vast majority of the craters on Earth and its Moon were created by asteroids but, thanks to the considerable gravitational pull of the giant gas planets, comets are more likely to strike bodies in these remote outposts.

Jupiter's moon Callisto is among the most heavily cratered objects in the solar system. Its fairly dark surface of ice mixed with dust is peppered with the bright scars of ancient craters. More than two hundred 10km (6.2 mile) features cover every million square kilometres of the cold surface. A clear difference between these craters and those on Earth's Moon for example is

Pages 92–93

Left: Saturn's moon Hyperion is viewed in spectacular detail by the Cassini spacecraft at a distance of 62,000km (38,500 miles). Hyperion is one of the strangest moons observed in the solar system, with clear evidence of extensive bombardment by meteors, and numerous landslides. Dark-floored craters dominate the 266km (165 mile) wide moon, with substantial differences in the composition of materials found in the craters.

Right: Saturn's moon Mimas has numerous impact craters but they are dwarfed by the Herschel crater seen in this false-colour image from Cassini. The crash that caused this 130km (80 mile) feature probably nearly shattered Mimas.

Previous page

This montage from separate images by the Galileo spacecraft compares the relative sizes of the asteroids Ida (above) and Gaspra (below). The surfaces of these rocky bodies are covered in numerous small craters, though it is interesting to note that Ida, which is 30km (18.6 miles) long, has had more impacts than Gaspra. This difference may be that Gaspra is younger and hasn't suffered as many collisions with other debris in space.

On 4 July 2005 an impactor probe from NASA's Deep Impact mission was deliberately manoeuvred to collide with the comet Tempel 1. The critical moment of this remarkable, staged event, is portrayed here in this artist's impression. The manner in which vast amounts of material are ejected from the comet was studied to provide valuable insights into the composition and structure of comets.

that Callisto's craters are not bowl shaped. Instead, the less rigid, icy surface of Callisto is able to deform, flatten, and restore the crust after an impact. The result is shallower and smoother craters. Another of Jupiter's moons, called Ganymede, has almost half its terrain covered in ancient craters, and has a similar appearance to Callisto's. Ganymede has a remarkable chain of craters, formed by a comet that broke into several pieces, which then struck the moon like shells from an automatic canon.

When it comes to impacts on a moon, the parent planet also plays a pivotal role, especially when the planet is a giant gas one. Saturn's immense gravitational pull attracts impactors that approach with increasingly great speeds. A consequence is that any moon located close to the planet is likely to experience far more frequent and energetic impacts than satellites much farther away. This is precisely the fate of Mimas, one of Saturn's innermost moons. Originally discovered by William Herschel in 1789, the moon has an icy and very heavily scarred surface with an average temperature of -200°C (-328°F). The terrain is so heavily cratered that any new impacts can only create marks over existing craters. One of Mimas's craters is remarkable for its large size compared with the moon itself. Named Herschel, this crater is about 130km (80 miles) wide, almost a third of the diameter of Mimas. A crater of similar relative proportions on Earth would be wider than Canada. The walls of Herschel crater are 5km (3 miles) high, though parts of the crater floor are 10km (6.2 miles) deep. The central peak of the crater looms 6km (4 miles) above its floor. The comet that struck Mimas billions of years ago probably almost destroyed it. Evidence of fractures can even be seen on the opposite hemisphere.

Striking the impactors

Recent spacecraft missions such as Galileo and NEAR have manoeuvred very close to asteroids and returned astoundingly detailed images showing that even the projectiles have craters. They have been observed on numerous asteroids, including Gaspra, Ida, Eros, and also Mars' asteroid-like moons Phobos and Deimos. The size and nature of these craters can provide clues as to the make-up of asteroids. Many craters are almost half the size of the asteroids. The asteroid Mathilde has dimensions of 66 x 48 x 46km (41 x 30 x 29 miles), but has at least five craters that are 19–33km (11–20 miles) across. They were probably formed by impacting bodies about 3km (2 miles) wide. If Mathilde were made of strong, rocky matter, it would have been destroyed by impacts of this magnitude. The fact that asteroids have such large craters indicates that they are mostly loosely glued piles of rubble (as opposed to solid rock). This looser composition allows the energy of impacts to be released close to the crater without spreading across the entire asteroid. This buffering is somewhat like a bullet, which will shatter a small rock, but be absorbed if it is fired into a pile of pebbles.

In July 2005, NASA scientists conducted an experiment to learn more about the chemical make-up of comets. After a six-month, 430 million km (267 million mile) journey from Earth, NASA's Deep Impact spacecraft approached a comet called Tempel 1. A small probe was released from the spacecraft and, in an impressive demonstration of space engineering, the probe was crashed into the comet with a relative speed of 37,000 km/h (23,000mph). The aim of the impact was to blast material from the comet's interior to be studied to reveal clues about the comet and the formation of the solar system. Several telescopes were trained onto Tempel to observe the impact, which launched debris into space and probably created a crater the size of a sports pitch. The data received is now being analyzed.

This dramatic image was taken 67 seconds after the Impactor probe deployed from the Deep Impact flyby craft crashed into comet Tempel 1. A large bright plume of dust was ejected upon impact, indicating that the comet's surface is covered in very fine, talcum-powder-like material. It is estimated that the engineered crash would have created a 100m (330ft) wide crater about 10m (33ft) deep on the comet's surface.

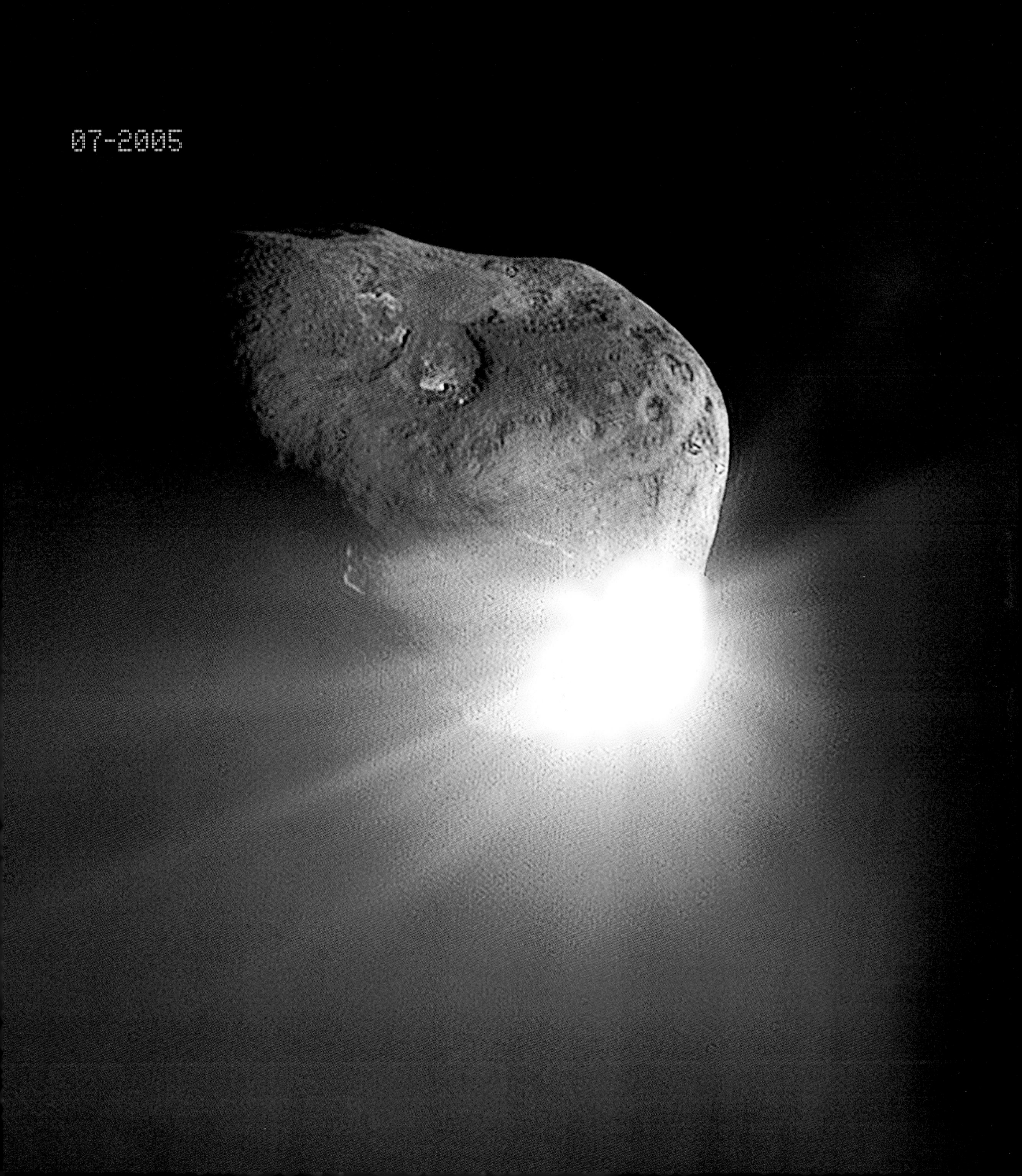
07-2005

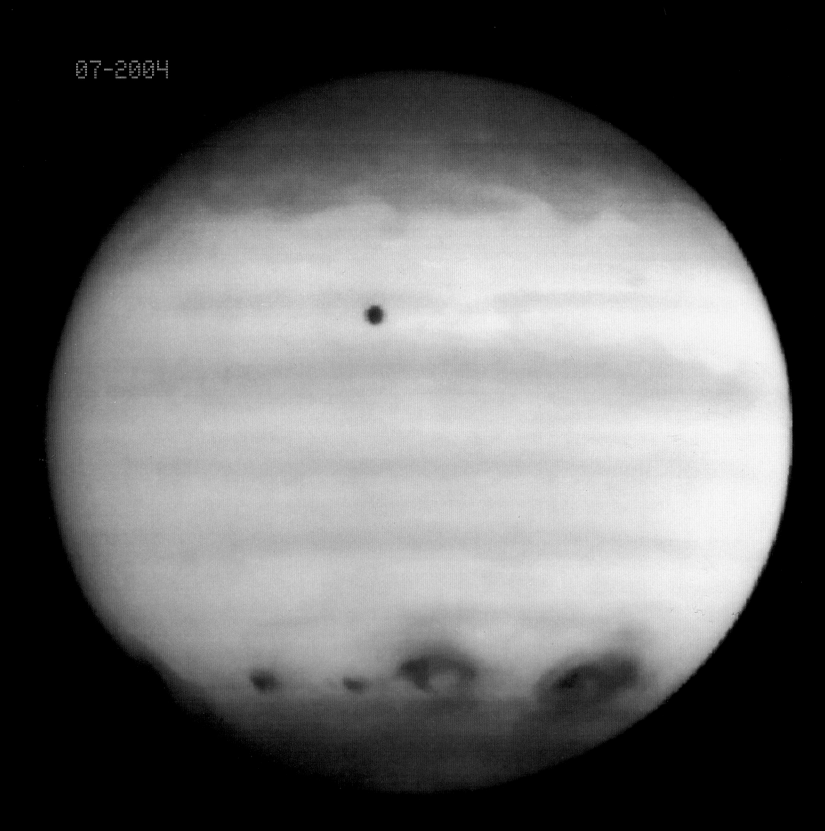

07-2004

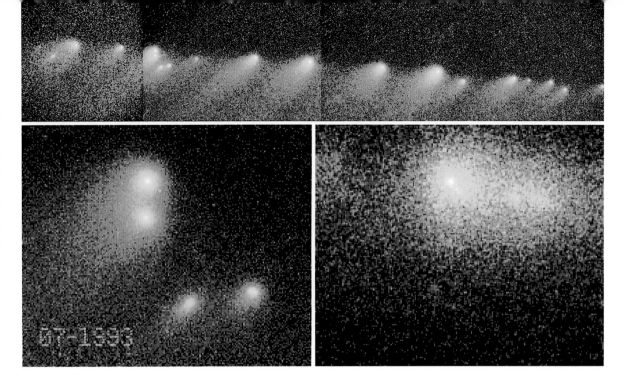

These images from the Hubble Space Telescope reveal that about a year before comet Shoemaker-Levy 9 struck Jupiter, the massive planet's gravitational pull had ripped the comet into more than 21 pieces. The largest of these fragments were 1–3km (0.5–1.9 miles) wide. This was the first time scientists were able to predict and then monitor a collision between two extra-terrestrial bodies.

Comet Crashes in Real Time

The aftermath of the crash of comet P/Shoemaker-Levy 9 into Jupiter is captured in this exquisite ultraviolet image from the Hubble Space Telescope. The huge fragments slammed into the upper atmosphere of the planet, leaving the dark circular patches in its southern hemisphere. The large quantity of dust from the pieces of comet absorbs the sunlight and thus appears as dark patches. The small dark spot above the centre in this image is Jupiter's moon Io.

So far in this chapter we have looked at the remarkable structures created by the bombardment of comets and asteroids that provide evidence of powerful impacts, primarily during the early history of the solar system. Because impacts are far less frequent today, it is very rare to watch a crash as it happens. In 1994, however, nature put on one of the most spectacular events ever witnessed by humanity. For the first time in history, the collision of a comet and a planet was predicted in advance and subsequently witnessed with powerful telescopes and spacecraft observatories. The collision in July 1994 of the comet P/Shoemaker-Levy 9 into the gas giant Jupiter would to advance our understanding of the phenomena of impacts. The composition of comets and the nature of the most massive planet in the solar system would also be studied during this awesome event.

The comet Shoemaker-Levy 9 (S-L 9 for short) was discovered in March 1994 by astronomers Gene and Carolyn Shoemaker and David Levy. It appeared squashed in the original images obtained from Mount Palomar observatory in the USA, but subsequent higher definition images revealed that S-L 9 had in fact split into several fragments. Within months the orbit of S-L 9 was established with enough accuracy to predict that the individual pieces were on a collision course with Jupiter; the great crash would happen in July 1994. Not surprisingly, the comet suddenly became the subject of intensive observations. Images from the Hubble Space Telescope revealed that S-L 9 had broken up into at least 21 pieces due to the stresses placed on it by Jupiter's immense gravitational pull. The size of the original comet was estimated to be about 8km (5 miles) wide, with the largest of the fragments possibly 1–2km (0.6–1.2 miles) across.

Between 16 and 22 July 1994, the 21 fragmented parts of the comet slammed into the back side of Jupiter as seen from Earth with speeds of 60km per second (37 miles per second). The impact region on Jupiter came into Earth's direct view about 30 minutes after each strike, carried around by the planet's rapid spin. The Galileo spacecraft had a direct view of the bombardment, a safe distance of 240 million km (150 million miles) from Jupiter. Fireballs of rising material were observed as the comet pieces plunged into the upper layers of the gas planet. The impact plumes rose thousands of miles above Jupiter's clouds and then collapsed back down. Fragments that were barely a few hundred yards across exploded with energies equivalent to millions of megatons. One of the headliners was a

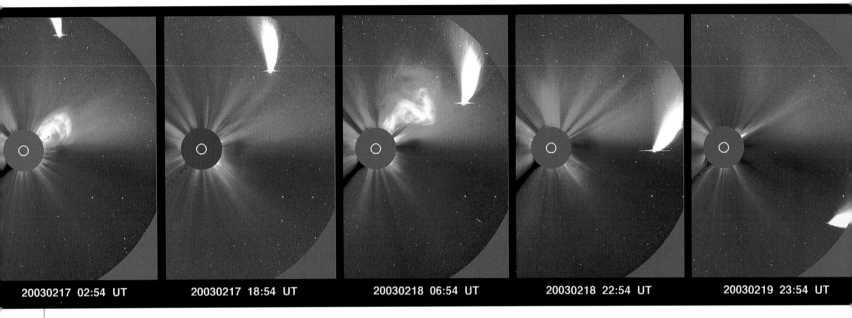

20030217 02:54 UT   20030217 18:54 UT   20030218 06:54 UT   20030218 22:54 UT   20030219 23:54 UT

These images from the SOHO spacecraft track the path of comet NEAT over three days as it approaches the Sun. The comet got to one-tenth of the Earth's distance from the Sun, and the intense heat boiled ice and dust from its nucleus. This time, it escaped the Sun and continued its journey to the outer regions of the solar system.

The detailed structure of an impact site on Jupiter's atmosphere from one of the huge fragments of Shoemaker-Levy 9. This false-colour Hubble image reveals a central dark spot about 2500km (1550 miles) across, surrounded by a "thin" ring almost 7500km (4660 miles) wide. The outer ring is nearly the size of the Earth.

fragment labelled "G" that blasted into Jupiter with an energy amounting to 6000 times the total estimated nuclear arsenal of the World. This single event created a 3000km (1865 mile) tall fireball above the Jovian cloud tops. This and other large disintegrations left Earth-sized scars in Jupiter's atmosphere and these markings persisted for several months to a year. The scars probably resulted from material in the comet itself, together with condensed portions of the plumes raised from the planet. Traces of silicon, magnesium, iron, and water vapour were all detected from the impact sites. The chemical mixture suggests that the comet pieces that vaporized were in fact loosely packed "dirty snowballs".

Kamikaze comets

Comets travel through the solar system in highly non-circular orbits that swing them around the Sun, followed by an outward journey that mostly takes them well beyond the planet Pluto.

They can take hundreds, thousands, or even millions of years to complete a single lap around the Sun. Sometimes, however, a few comets get a little too close to the Sun and plunge into the solar atmosphere. Since the late 1990s a spacecraft called the Solar and Heliospheric Observatory (SOHO) has been studying the Sun and the environment around it. Over this relatively brief period SOHO has discovered more than 700 new comets. A surprisingly large number of these are suicidal comets that vaporize and get dragged into the Sun. All of these icy bodies are of course insignificant in size compared with the Sun and they have no effect on our backyard star.

The fate of these SOHO-discovered comets is similar to that of comet Shoemaker-Levy 9, only this time it's the immense gravity of the Sun that tears the comets apart. The study of the kamikaze comets is important in teaching us about their make-up and strength, which could provide vital information for our own protection on Earth.

Watch the Skies

We have seen in this chapter that ever since its formation, the planet Earth has been bombarded by comets and asteroids. This barrage of projectiles has reshaped the surface of our planet and occasionally triggered the extinction of living species. Ironically, incoming bodies have also carried water and carbon to the Earth, which are essential ingredients for life. While the Shoemaker-Levy 9 and SOHO comet crashes have enthralled scientists in recent years, they also remind us that 4.5 billion years after its formation, the solar system is still a dangerous place. These powerful strikes have raised new questions about the possibility of similar impacts on Earth.

The primary threat to Earth today comes from Near-Earth Objects (NEOs), which are asteroids and comets on short orbits around the Sun that bring them close to Earth. It is estimated that there may be 1100 NEOs of diameters exceeding 1km (0.6 miles) and 100,000 that are about 100m (330ft) wide. These are approximate figures, however, and they could easily be underestimates. International teams of astronomers are constantly surveying the skies to detect and calculate the precise orbits of the potentially Earth-approaching hazards. It's a huge task that requires dedicated telescope facilities, such as the Spacewatch survey in Arizona, USA, and the LINEAR search programme in New Mexico, USA. The Space-watch project uses a modest 90cm (35in) telescope to look for small objects over a limited patch of the sky. Using two automated telescopes, the LINEAR team has discovered more Near-Earth Objects than any other group. The aim of these groups is to discover and chart the motions of 90 percent of the vagabond bodies that are more than 1km (0.6 miles) in size.

So what might be the potential environmental effects on Earth from an impacting Near-Earth Object? Thankfully, the Earth's atmosphere acts to shield us from most of them. Impactors less than 50m (165ft) in diameter burn up or explode high in the atmosphere. The consequences of these events vary from loud bangs and a shower of fragments, to airbursts that are far more powerful than hurricanes. In 1908 a 50m (165ft) asteroid-like object resulted in an airburst at Tunguska in Siberia, which released the equivalent energy of 10 megatons, flattening almost 2000 sq km (770 sq miles) of forest. The greater the size of an incoming projectile, the greater the energy it releases, and thus the greater its potential threat. The precise outcomes of a strike on Earth are hard to predict because they depend on several factors, including in particular the make-up of the object. For example, Near-Earth Objects about 150m (770ft) across would destroy city-sized areas on land and create vast tsunami waves in oceans. Objects a few miles across would yield up to 10 million tons of equivalent energy and potentially devastate areas the size of large nations. Impacts from objects over 10km (6.2 miles) wide would cause global disruption, with large-scale extinction of living organisms. Mercifully, such events are very rare. Near-Earth Objects that are a few miles in size are estimated to strike Earth once every million years or so, while impacts posing global threats average one every 50–100 million years. In terms of frequency, smaller NEOs between a few hundred metres/yards to 1km (0.6 mile) are possibly more dangerous because they may strike every 50,000 years and still cause severe, if local, damage.

There is a growing realization that, at minimum, Near-Earth Objects need to be monitored and space missions such as Deep Impact are important for advancing our understanding of the structure and composition of asteroids and comets. In September 2005 the Hayabusa spacecraft of Japan's Space Agency met up with an asteroid named Itokawa, 110 million km (180 million miles) from Earth. The spacecraft is attempting to collect a sample of material from the surface of the asteroid and bring it back to Earth in June 2007. If the mission is successful, Hayabusa would be the first spacecraft ever to return samples from an asteroid. Scientific analysis of materials from the asteroid's surface are expected to provide pivotal information about how these bodies were assembled billions of years ago.

At the time of writing, none of the Near-Earth Objects with known orbits is predicted with certainty to be on a collision course with Earth. However, the fact that a large number of objects remain to be tracked, and many others await discovery, makes it very difficult to predict the impact threat. It is clearly important that we should continue to catalogue large objects, because our chances of mitigating the effects of a strike largely depend on how much warning we have. Several times a year modest-sized asteroids sail pass the Earth unnoticed and unpredicted. Recent near-misses include the object named 2002 MN that buzzed our planet at a distance of only a few tens of thousands of miles; this is very close by the dimensions of the solar system. Another asteroid designated 2001 YB5 passed the Earth in January 2002 at a distance of 805,000km (500,000 miles), which is less than twice the distance to our Moon. The object is thought to be hundreds of metres across, and it carried the potential for substantial destruction. Alarm has also recently been raised by the discovery of an asteroid originally designated 2004 MN, which is in an Earth-like orbit around the Sun. Current predictions of its orbit are that it will pass within just 36,000km (22,350 miles) of Earth on 13 April 2029. At such a short distance, the asteroid will be seen with the naked eye as it streaks rapidly across the sky. It is obviously critical to refine the orbit of the object, and make the most accurate predictions of

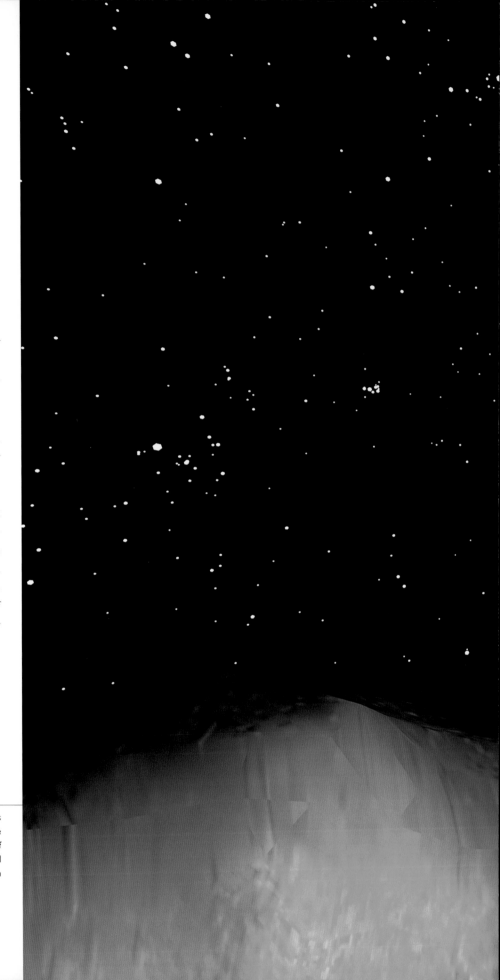

its path. With an estimated diameter of 300m (985ft), the asteroid carries a potential impact energy of at least 1000 megatons and is thus capable of causing severe devastation. Indeed, 2004 MN has recently been renamed 99942 Apophis, after the Egyptian god of destruction.

The hope is that the current and future surveys will provide years, or perhaps decades, of advance warning of a NEO on a collision course with Earth. In this scenario it would be realistic to consider deflecting or destroying the object. For example, several spacecraft could be launched to intercept the object, and explosives used to nudge it onto a different orbit well away from Earth. A major uncertainty for any mitigating mechanism is that we simply do not know enough about the internal structures of asteroids and comets. If they are indeed loose piles of rubble held together by their own weak gravitational attraction, then any effort to deflect Near-Earth Objects may result in their fragmenting into hundreds of objects. The consequence then could be an even greater danger, over a considerably larger area. In terms of risk assessment and mitigation, new spacecraft missions designed to rendezvous with Near-Earth Objects offer the best prospects of refining techniques and conducting in-situ investigations. The European Space Agency is, for example, studying designs for a mission called Don Quijote. The concept is to launch two spacecraft, one that will impact a small pre-selected asteroid at very high speeds and one that will orbit the same asteroid for months before and after the impact. The idea is to monitor changes in the asteroid's internal structure, spin rate, shape, and orbit caused by the impacting craft. This type of mission represents an important step in testing the technology required for future attempts to deflect objects in space.

Rocky remnants in space pose a potential natural threat to our planet, as ominously depicted in this artist's impression. Current efforts to mitigate against any strike from Near-Earth Object involve telescopic monitoring of the skies to determine precise orbital positions of rogue rocky bodies, and launching spacecraft to rendezvous with asteroids and comets to learn more about their physical make-up.

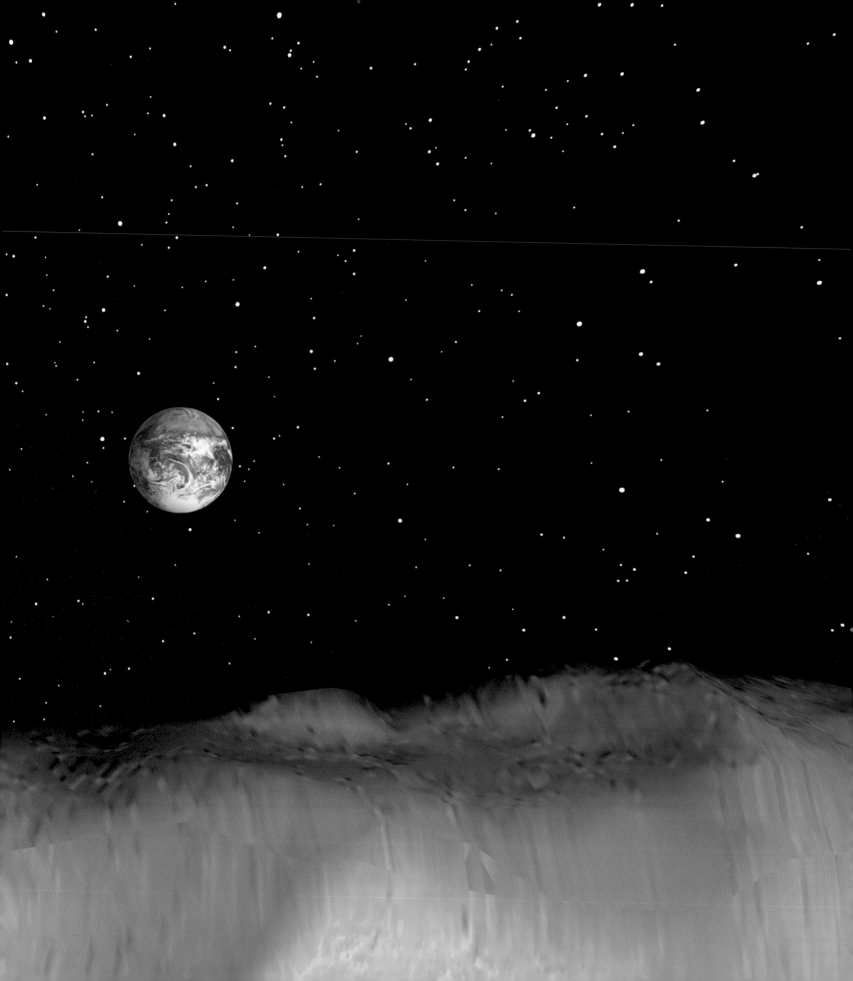

In an unimaginably vast Universe, all the living organisms that we know are on our own planet, contained within in a boundary marked here by the Space Shuttle *Endeavour*, at an altitude of 300km (185 miles). A major challenge in the modern exploration of the solar system is to to look for evidence of life beyond Earth.

The Search
for Life

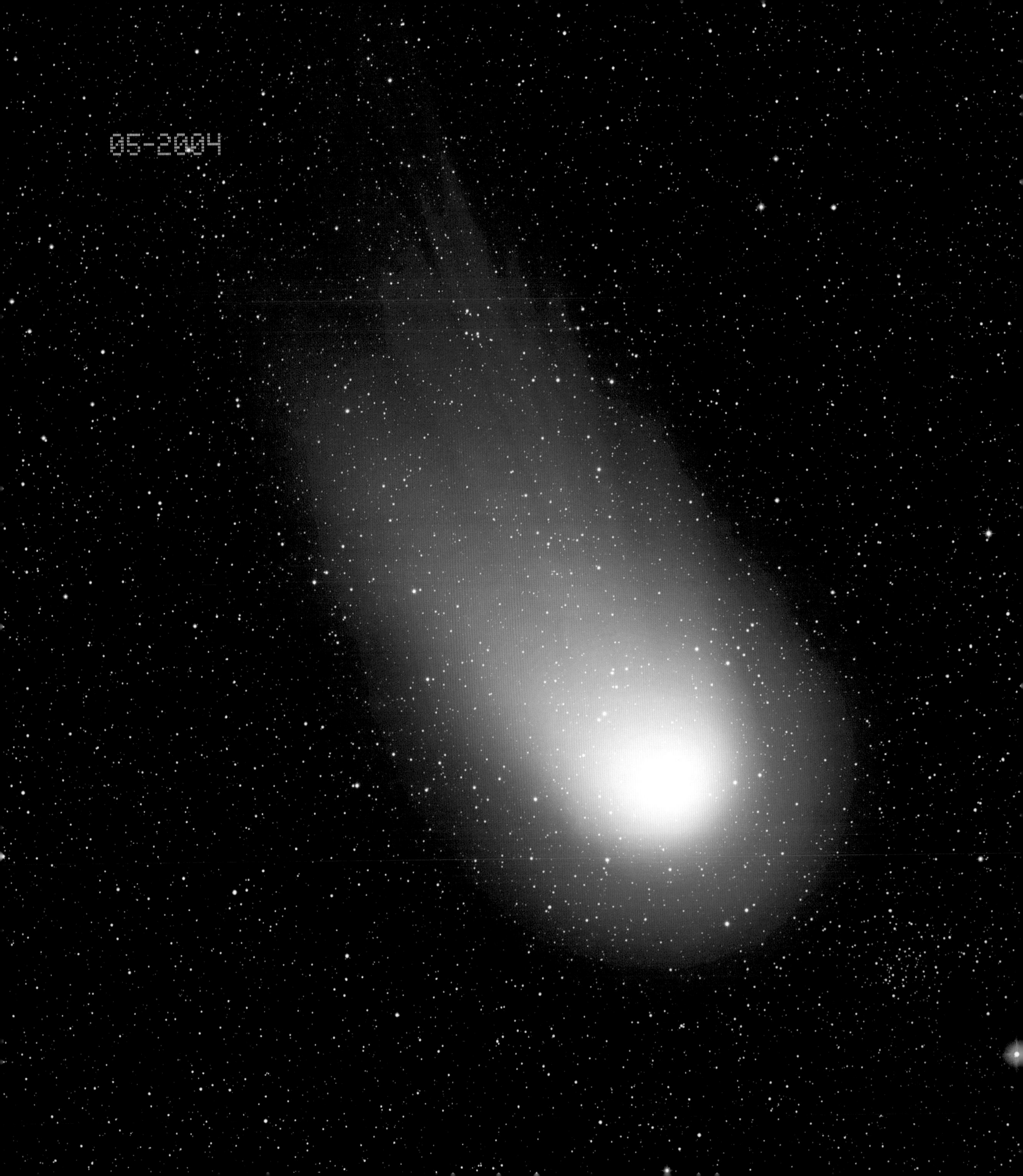
05-2004

One of the great quests in the 21st-century exploration of the solar system is to understand, and search out, the prospects for life beyond Earth. Although the question of whether or not we are alone in the Universe dates back to ancient times, never has this profound issue been more actively studied. The discovery of primitive microbial life-forms thriving in extreme environments on Earth has motivated the search for life in similarly harsh conditions elsewhere in the solar system. We are living in a dramatic "golden age" in which the sciences of astronomy, biology, chemistry, and geology, are brought together under the umbrella of "astrobiology" in the pursuit of extra-terrestrial life. The hunt is for nothing more sophisticated than primitive microbes and bacteria, but nevertheless their discovery would be most profound in considering our place in the Universe. This chapter looks at the role of water as a necessary medium for life and asks what biosignatures would betray life on other worlds. Our tour of potentially habitable environments for bacterial life takes in the surface and subterrains of Mars, the predicted oceans of liquid water under the ice-encrusted moons of Jupiter, and the latest discoveries from the Cassini mission to Saturn's rich moon, Titan. The identification of potential habitats poses technological challenges for studying conditions and collecting evidence. Understanding the relationship between the occurrence of life and the evolution of planets will ultimately set us on the path to seeking out biological processes in planetary systems around other stars.

The impressive sight from Kitt Peak National Observatory of the comet NEAT streaking into the inner solar system. Comets have played a key role in delivering water ice and complex molecules onto the surfaces of planets and moons throughout the history of the solar system.

Life on Earth

In searching for a strategy to explore life in the solar system, it makes sense start with at the one example we know – life on Earth. It is possible that life elsewhere may have developed in different ways to life on our planet, but we have to base our pursuit on our current knowledge of life and the corresponding habitable environments. We should also acknowledge from the outset that any life-forms discovered in the solar system beyond Earth will be primitive – microbes and bacteria. Microbial life has thrived on Earth for more than two billion years and it is spread across virtually every environment. The extensive imagery now gathered of surfaces and conditions on other planets and moons suggests that it is very unlikely that more sophisicated life exists anywhere else in the solar system. Meanwhile, the characteristics of evolution and reproduction are shared by most life forms on Earth and so provide a basis for exploring primitive life in the solar system. Requisites include a liquid medium such as water, sources of energy, and an abundance of the nutrients and chemicals needed to build living cells.

Life at the extremes

In recent years our understanding of the diverse conditions in which life thrives on Earth has provided vital new perspectives on the prospects of life elsewhere. There are life forms on Earth that appear to be able to survive extremes of temperature, or in environments lacking in water, oxygen, or light. Organisms that can withstand harsh conditions are called extremophiles. There is a wide variety of these organisms that are highly adaptable to pressure, acidity, freezing, dryness, radiation, and so on. These extremophile microbes provide terrestrial examples of possible extra-terrestrial life and exobiology. Specific examples of extremophiles include tiny multi-cellular organisms called tardi-grades that can survive temperature ranges of -130°C to +130°C (-200°F to -260°F). A form of bacterial life called Archaea can thrive inside volcanoes, or even in deep-buried rocks on Earth.

Equally remarkable are organisms that live in deep-sea volcanic vents, where temperatures exceed several hundred degrees in a toxic environment of metals. The vents sit over regions of molten rock, which heat the water and help to dissolve a range of chemicals, including mercury and gold. As the heated water cools, the chemicals precipitate to create black "smoke" that is commonly associated with hydrothermal vents. Life forms, such as mussels, which surround these regions essentially feed on this metallic smoke. These deep-sea chemical reactions also provide energy for the growth of bacteria. Tenacious microbes seem to adore the volcanically active regions of our planet. We will see shortly that the habitats offered by the hydrothermal vents in Earth's seas may be

There is no place like home. At the start of its journey to the innermost planet, Mercury, the Messenger spacecraft took this stunning image of our planet, Earth. No other body in the solar system has an atmosphere in which humans can breathe, or the great volumes of surface water that we can drink. Millions of different species of plants, animals, and micro-organisms thrive on our remarkable planet.

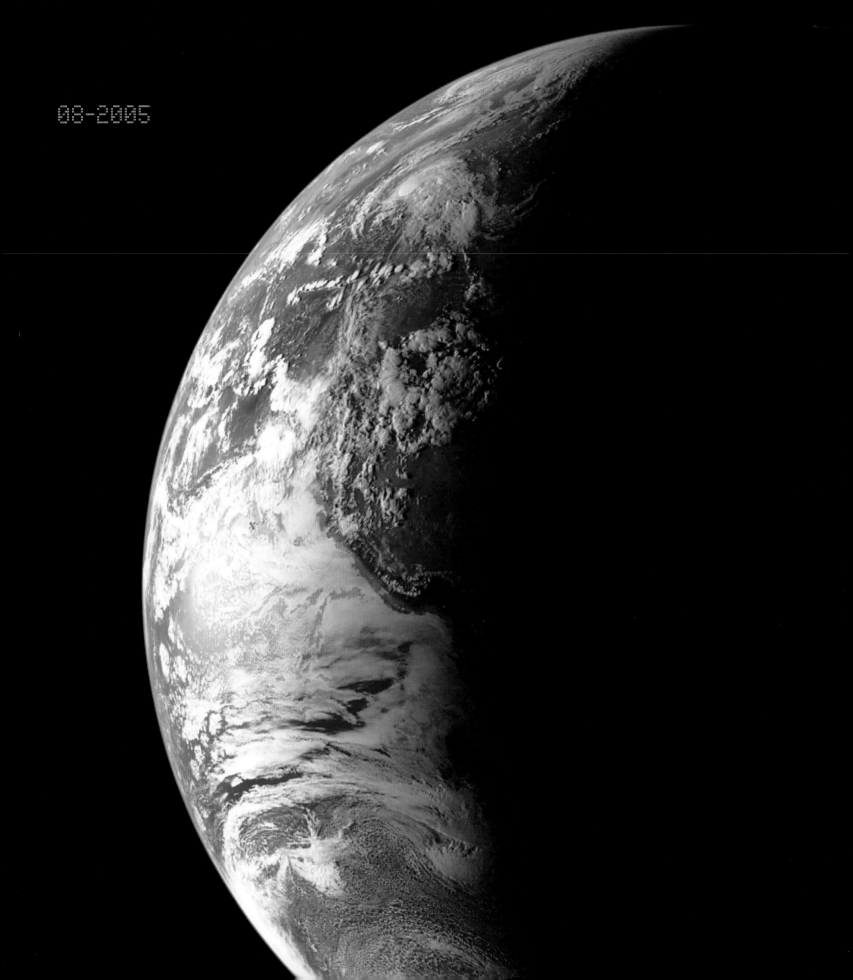

similar to conditions beneath the vast oceans thought to exist on Jupiter's moons, such as Europa (see pages 128–31). Another extreme environment on our planet is the sea-ice that covers the Arctic ocean. This ocean is composed of water ice, minerals, salts, and pockets of air. Despite drops in temperatures to less than -25°C (-13°F), the Arctic ice is colonized by micro-organisms that exist beneath the frozen ice shelves, even during long and dark winters. Our understanding of these how these organisms survive is very important in the context of the moons of the giant gas planets, which are frequently encrusted in solid and slushy ice. Equally extraordinary are the microbes that live several miles deep beneath the Earth's rocky surface. Despite a lack of sunlight and carbon, they survive in the water that flows through cracks in the surface, gathering nutrients as it passes through layers of basalt and granite. This sub-surface biomass on our planet provides us with a fascinating comparison with the wet environments thought to exist below the dry crust of the planet Mars.

The big deal about water

The Earth seems to be unique in the solar system in that there is an unambiguous link between the presence of vast volumes of liquid water and a spectacular variety of living species. It is therefore reasonable to speculate that extra-terrestrial life in the solar system might also require liquid water to evolve and survive. The chemistry of water, and the way it changes with temperature, makes it an ideal medium for biology. Water has also played a dominant role in the geology and climate of our planet. It was

The rich colours of the Grand Prismatic Spring in Yellowstone National Park, USA, are due to the presence of different species of heat-loving bacteria called thermophiles. The bacteria thrive in the temperature range between the hot central regions of the pool and the cooler waters along its edges. The red pigment-producing bacteria survive in the cooler waters, while yellow and green pigments trace bacteria living in progressively hotter water.

the pivotal habitat in which life first appeared on Earth billions of years ago, and it provides biologically useful minerals from Earth's crust and mantle. On the smaller scale of living cells, liquid water dissolves organic material and transports chemicals to make them useful for metabolic reactions within cells.

Another advantage that water has over, say, methane and ammonia, is that it remains in liquid form over a wide range of temperatures, which is of great benefit when faced with severe conditions or climate changes. Even in extreme cold, when the temperatures are low enough for water to freeze, water ice is less dense than liquid water, so it floats to form a layer above seas and lakes. This icy covering then acts like a blanket, allowing the water beneath it to remain liquid, thus enabling the sea life within it to survive.

Before we go on a hunt for ice and liquid water in the solar system, it's worth noting that, besides water and nutrients, life on Earth also needs energy. The process of photosynthesis that is so dominant in the Earth's ecosystems is reliant on light from the Sun. The heat from our planet's internal molten layers is, however, of greater relevance in the context of distant planetary bodies that receive very little sunlight. We have already seen that the Earth's internal heat permits bacteria in hydrothermal vents on the ocean floors to make food. This process, known as chemosynthesis, is significant for other planets and moons within the solar system, because it is totally independent of the amount of sunlight.

Looking for life farther afield

As you can see, lifeforms on Earth can be very opportunistic and can establish a foothold in extremely hostile environments. Our realization of this has led to a shift in assessing conditions in the solar system that might host alien microbes. The most significant point is that life on Earth doesn't need light, so life could also flourish beneath the ice crusts of other planets or moons. Let's take a tour of locations in solar system that may provide favourable conditions for sub-surface life, starting with Mars, and then looking at the icy moons in the outer solar system.

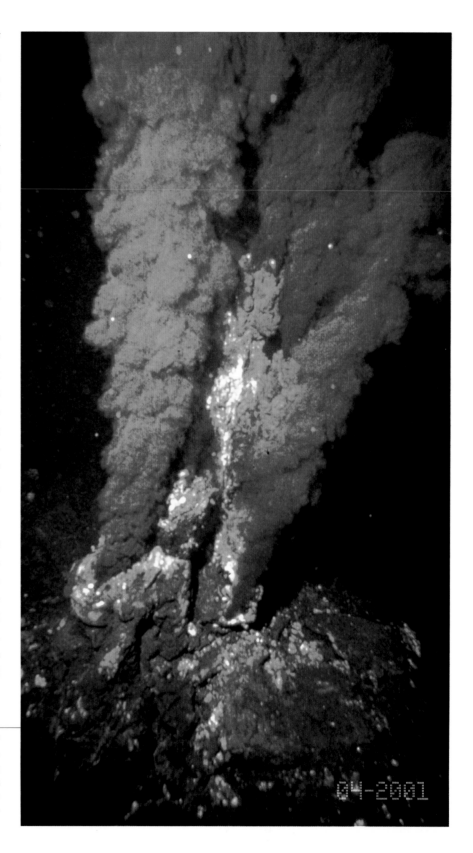

Black smoke spewing up from a hydrothermal vent at an ocean depth of more than 2200m (7200ft), where volcanic activity is associated with an expansion of the sea floor. The structures are composed of sulphur-bearing minerals that originate beneath the Earth's crust. These hydrothermal vents provide the energy to support exotic life forms and entire ecosystems on Earth, without the need for sunlight. Life could possibly have formed in similar environments on other bodies in the solar system.

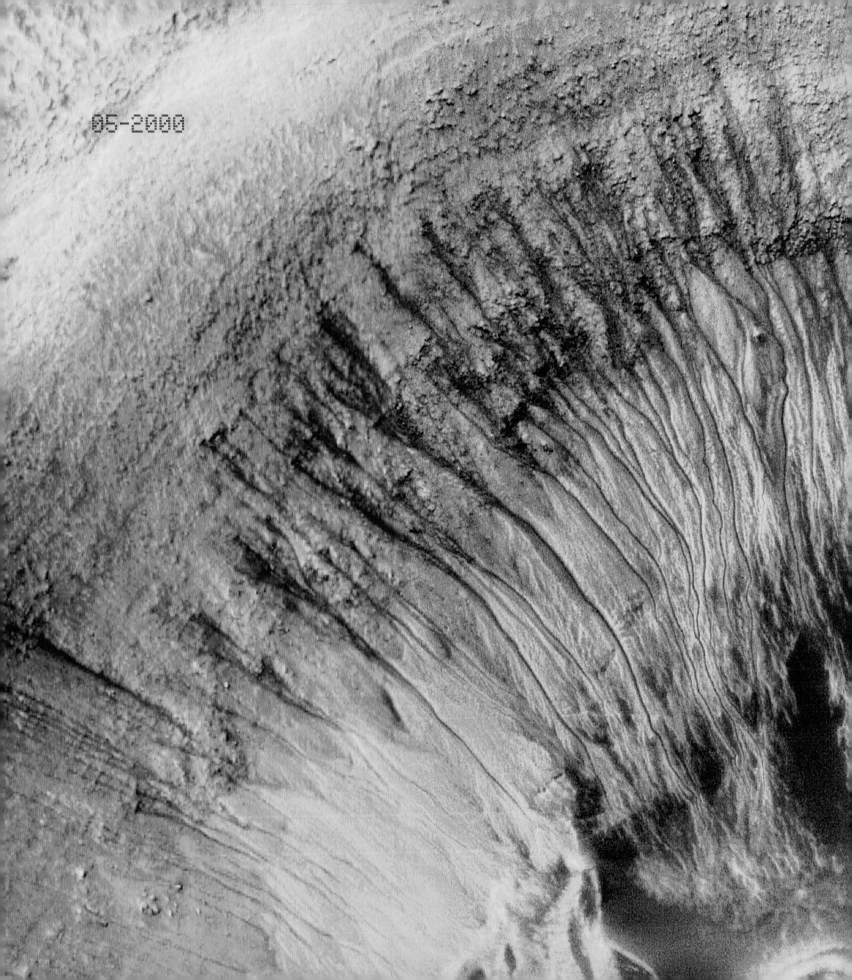

Life on Mars?

Spacecraft exploration of our neighbour Mars has ruled out previous fantasies and fictitious speculation about intelligent beings and sophisticated cities on the red planet. Nevertheless, Mars is still a target of intensive study because it probably represents our best prospect of finding evidence for past or present microbial life beyond Earth. Close-up images of the Martian surface have revealed unambiguous evidence that substantial amounts of water flowed there billions of years ago. Examples include massive channels carved by running water, ancient river beds, flood plains, and layers of sedimentary rock. Some of the large, long-extinct, volcanoes reveal water-carved features in their slopes, which point to a connection between ancient floods and volcanic heating. Also intriguing is the unconfirmed possibility that a vast ocean may once have filled low-lying basins near the northern polar regions.

Dramatic climate change on Mars

Somewhere around three billion years ago the climate of Mars changed drastically to transform it from a wet and warm planet to the cold, dry, and desert-like body we see today. Though liquid water is no longer on the surface, water in the form of ice is certainly present and may even pervade beneath the surface.

Previous page
This unusual perspective of the Grand Canyon in Arizona, USA, was provided by the Ikonos Earth-monitoring satellite. Hundreds of miles long and up to 30km (18 miles) wide in places, this is the largest canyon on Earth. Layers of sedimentary rock have been exposed over millions of years by the action of moving water, the key to life. The sediments and rock layers of the Grand Canyon record billions of years of Earth's history.

The wall of this small impact crater on Mars has numerous narrow gullies carved into it. The clearly visible channels in this image from the Mars Global Surveyor are thought to be the result of erosion due to the flow of water and other debris. Hundreds of flow events are on view here, with an estimated 2.5 million litres (660,000 gallons) of water flowing in each one. The relatively crisp edges of these gullies suggest that the water flows occurred in geologically recent times and could still be present on Mars today.

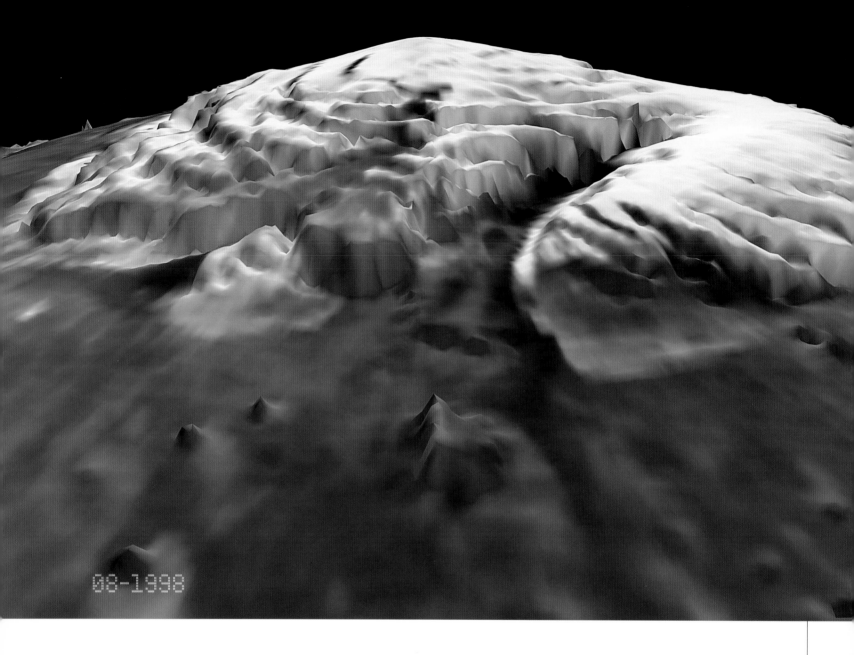

08-1993

The polar caps of Mars

Since the earliest telescope observations from Earth, astronomers have noted that Mars has polar caps on its north and south poles. Today our understanding of these regions is being advanced by detailed monitoring from sophisticated orbiting spacecraft such as NASA's Global Surveyor and the European Space Agency's Mars Express. Although Mars is still a mystery, the current view is that the northern ice cap is mostly frozen water, while Mars' southern cap is a mixture of frozen water and carbon dioxide ("dry ice"). No one is exactly sure how much water ice there is in Mars' polar caps, but its presence is pivotal to understanding the history of the planet, and as a potentially useful resource for future human exploration on the planet.

As on our planet Earth, the two Martian ice caps are very different. On Earth, the northern cap is permanent sea ice, while the southern cap is above solid rock (the continent of Antarctica). On Mars, the northern ice cap appears to be pitted ice over rocky plains. The southern cap is more disparate. It is 6km (3.7 miles) higher than the northern region, with complex and strange shapes etched into it by erosion. These shapes form spiral depressions that curl around the pole, revealing internal layers of the icy region. Another similarity is that the polar regions of Mars expand in the winter and shrink in the summer.

This dramatic image captures Mars' icy northern polar cap. The unique three-dimensional representation of the pole was obtained using laser measurements with the Mars Global Surveyor. This ice cap is 1200km (745 miles) across and up to 3km (1.8 miles) thick. It is scarred by canyons and troughs up to 1km (0.6 mile) deep.

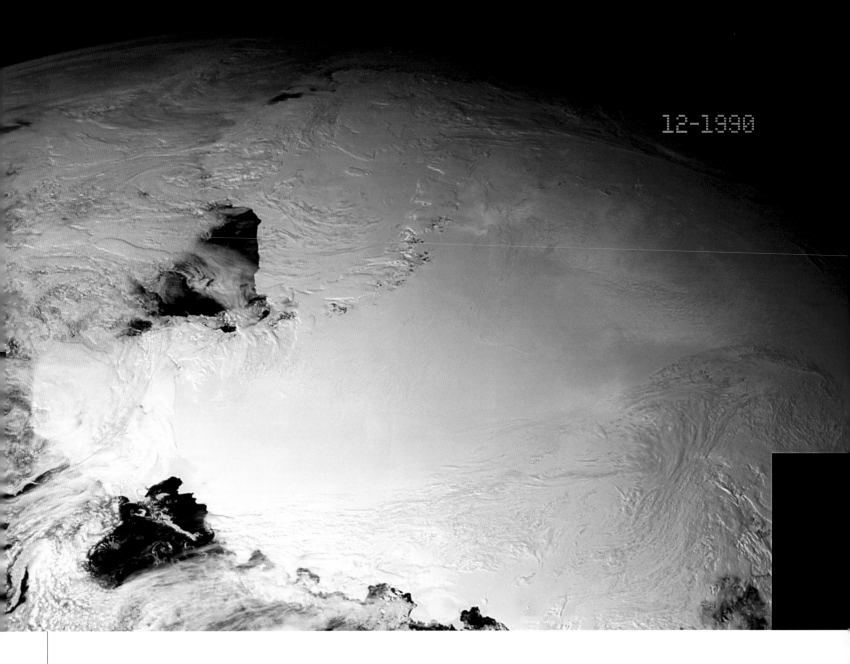

12-1999

In comparison, this image shows Earth's southern polar cap, Antarctica, viewed by the Galileo spacecraft en route to the planet Jupiter. The Antarctic ice is up to 5km (3 miles) thick, and spans 14 million square km (5.4 million square miles). This southern summer solstice scene is dominated by the vast Ross ice shelf.

Like Earth, Mars has a cycle of seasons, and its axis of rotation is tilted by a similar amount to that of Earth (about 25 degrees). The big differences, though, are that Mars takes about two years to orbit the Sun, so it has longer seasons, and it has a much thinner atmosphere, so it is poorly insulated against temperature fluctuations that occur as it orbits the Sun. Another important factor compared with Earth is that Mars' orbit is less circular, so its distance from the Sun changes more significantly over the course of its year.

When Mars is closest to the Sun and it is summer in the south pole, it basks in 40 percent more sunlight than it gets during winter. At maximum size the southern seasonal cap is almost 4000km (2485 miles) across; half a year later the northern cap is at its largest with a diameter of 3000km (1865 miles). While the Martian seasonal caps grow very large and then shrink, Mars also has small permanent caps, about 1000km (620 miles) across in the north and 350km (220 miles) wide in the south.

Martian exploration

Our exploration of the potential water warehouses on Mars is in its infancy. In December 1999, NASA's Polar Lander spacecraft was due to touch down on Mars' southern ice cap to provide the first in-situ analysis of samples of the polar ice. Sadly,

contact with the spacecraft was lost prior to landing; the cause remains unknown. Meanwhile, the polar caps of Mars remain fascinating sites for further exploration. Sub-surface microbes could survive in the polar ground during the summer and lie dormant during the winter. The lessons from Earth's Arctic suggest that polar regions permanently covered by ice could provide a vital insulation for sub-surface lakes below, where primitive life may survive.

Since 1997 and the arrival of the Mars Global Surveyor spacecraft at Mars, new images have provided remarkable evidence that water has flowed there in more recent times, and possibly still does so in localized regions. The latest views reveal relatively fresh gullies with sharp (not heavily eroded) edges that contrast against the walls of craters. Scientists believe that the gullies may be the result of water released in small underground flash floods. It is exciting to think that these gullies may still be forming today and could provide a current habitat for living organisms to survive. A fascinating analogy on Earth to what may be awaiting discovery beneath the Martian surface has recently been uncovered deep below the Beverhead Mountains in Idaho, USA. Scientists from the US Geological Survey have discovered a strange species of micro-organism. Besides demonstrating that microbes can survive without sunlight, it appears that these deeply buried organisms are harnessing hydrogen gas that was released in our planet's interior. The findings in Idaho raise the possibility that hydrogen-based life may thrive below the surface of Mars. The notion is that the micro-organisms rely on combining hydrogen with carbon dioxide to yield methane gas. This process, which does not involve organic carbon, is precisely what is mooted to be a possibility on Mars.

Any serious attempt to discover microbial life on Mars would of course have first to identify the most strategic places to look. Following the discoveries on Earth, one possibility is to explore geological features that betray the long-term presence of water. There is certainly no shortage of space missions probing Mars. Between 2001 and 2005, the Mars Global Surveyor, Mars Odyssey, and Mars Express spacecraft orbited and monitored the planet, while the twin Mars Exploration rovers successfully travelled across the surface, examining the terrain and variety of rocks. Over the next few years further spacecraft should arrive, such as the Mars Reconnaissance orbiter and more "smart" landers and rovers. Some of the new

missions orbiting Mars will seek to use infrared sensors to map the spread of minerals on the surface, including silicate- rich and coarse-ground rust that may have formed in ancient shallow seas. Infrared data can also help to locate warmer regions of Mars that could even today host warm springs or evaporation. The collective data from current and future missions will help pinpoint ideal exploration sites, which could be targeted for new robotic landers, missions designed to bring rock samples back to Earth, and perhaps even missions sending teams of humans. A realistic possibility for the end of this decade is the Mars Science Laboratory planned by NASA. Far more ambitious than the Exploration rovers currently touring the Martian surface, this new vehicle aims to collect soil and rock samples, which would be analysed for traces of organic compounds, such as proteins and amino acids. The new mission would also aim to examine the composition of the atmosphere for subtle signs that might indicate biological activity. The intension is that the Mars Science Laboratory would be uniquely steered to a desired location on Mars, in a similar way to the Space Shuttle's entry into the Earth's atmosphere. The use of parachutes before landing should allow the roving laboratory to land within an area 20–40 km (12–24 miles) across.

Two outflow channels named Dao (to the north) and Niger (south) are shown in this image from the the Mars Express spacecraft. Both structures are about 40km (25 miles) wide and 1400–2400m (4590–7,870ft) deep. These eroded valleys are carved on the flanks of a volcano and their shapes indicate that they were formed when sub-surface water seeped out, resulting in the collapse of the surface ground layers.

Previous page
This spectacular feature on the surface of Mars shows a vast patch of water ice on the floor of a crater. The intriguing view was obtained using the high-resolution stereo camera on board the European Space Agency's Mars Express spacecraft, which delivers colours that are close to natural. Traces of water ice can also be seen along the crater's rim and walls. The crater itself is 35km (22 miles) wide, and about 2km (1.2 miles) deep.

Overleaf
The sharp camera on board the Mars Express spacecraft zoomed in to obtain this detailed view of the floor of a Martian crater called Nicholson. The remarkable scene shows a structure that rises 3.5km (2.2 miles) from the central floor of the 100km (62 mile) wide crater. There is evidence of substantial erosion in the flanks of this peak, with some of the sculpted markings probably caused by water flowing in ancient times.

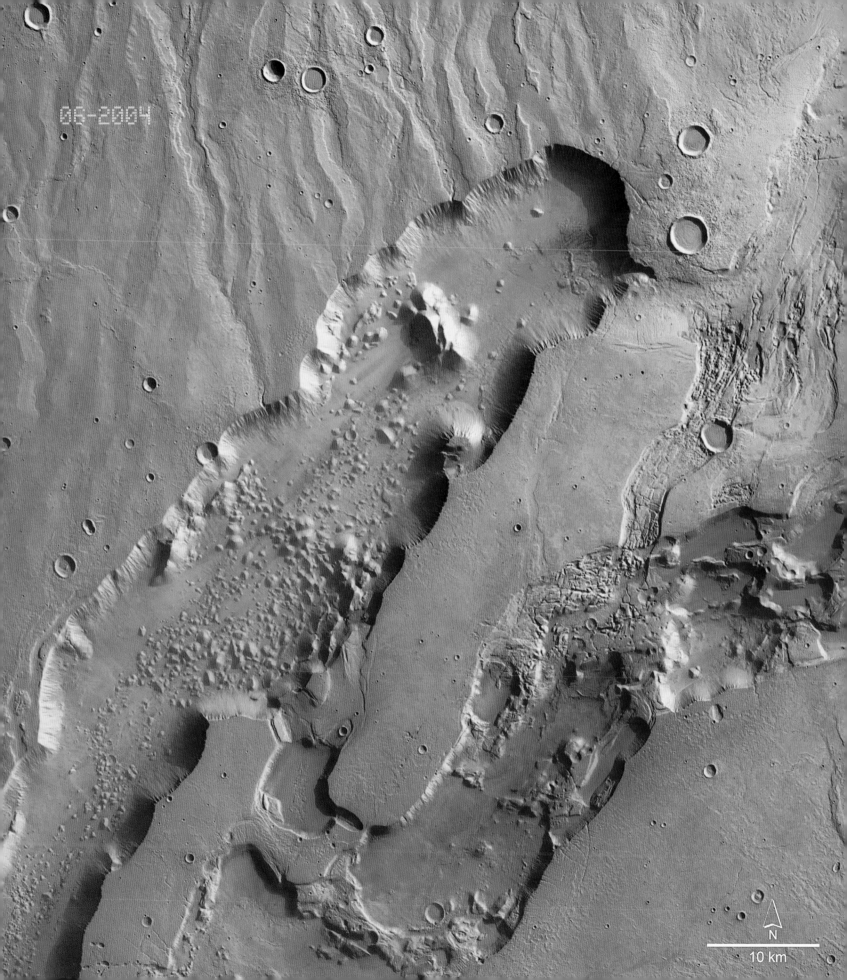

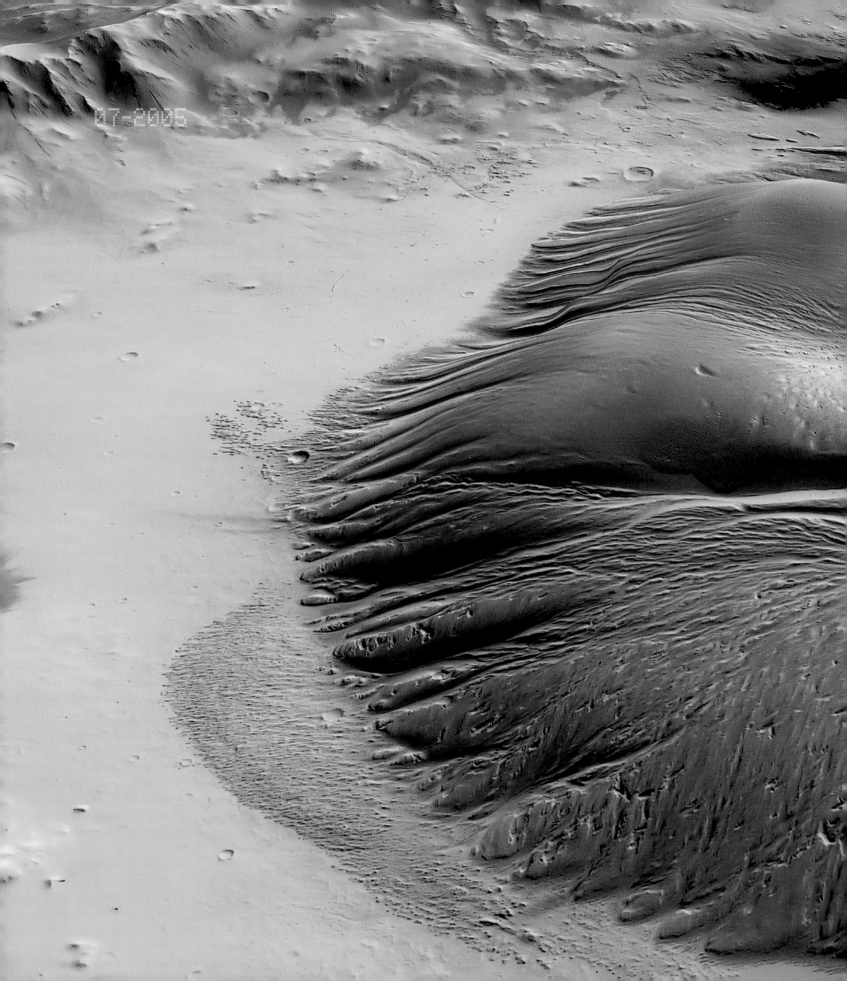

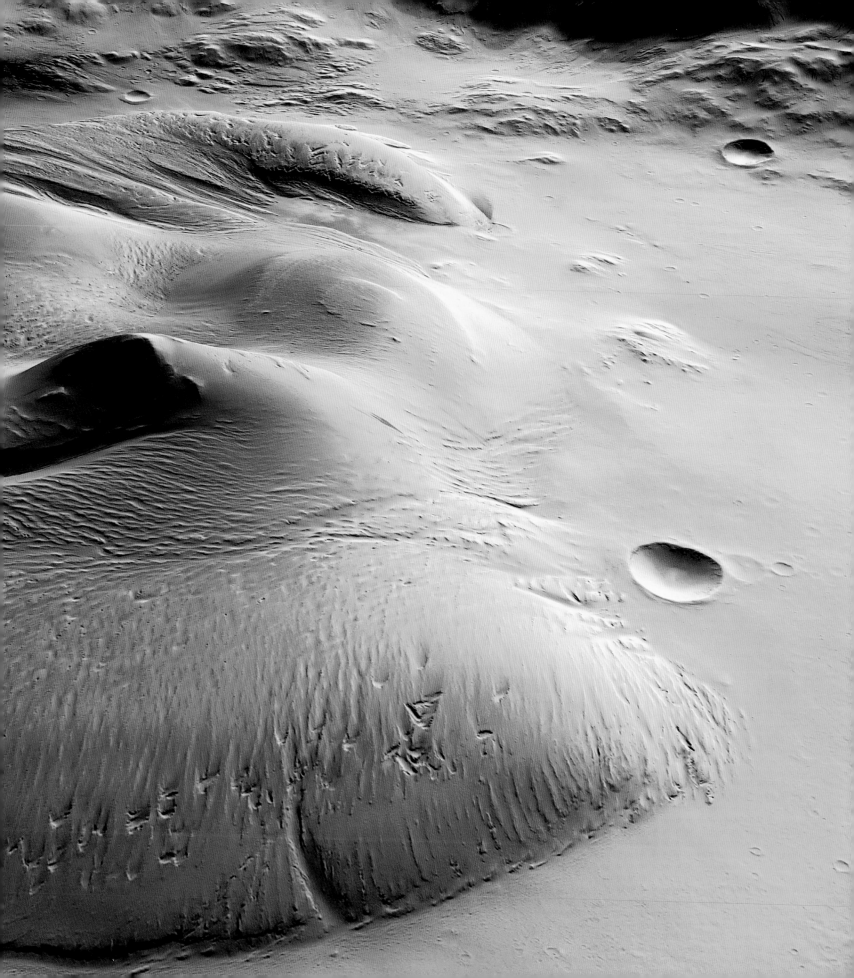

Ice-encrusted Moons

Until recently, the notion of finding life in the freezing outer regions of the solar system seemed outrageous. The latest explorations of the ice-covered moons of Jupiter and Saturn have, however, raised remarkable new possibilities for these locations as habitats for life. What all of these icy bodies have in common is the prospect of dark ecosystems beneath a frozen crust, which may be warmed by internal heat. The notion is that microbial communities could in fact thrive deep in melted ice, with access to chemicals and nutrients.

Jupiter's icy moons

Between 1995 and 2003, NASA's Galileo mission beamed back highly detailed images of the surfaces of three of Jupiter's large moons, Europa, Ganymede, and Callisto. These views, together with other data and measurements, transformed our knowledge of these remote worlds and provided evidence that their surface ice may be overlying oceans of salty, liquid water. Europa in particular, has a relatively young surface, and images from Galileo reveal a surprising lack of impact craters for a body that is not protected by an atmosphere and is so close to a comet-attracting giant planet. The suspicion is that the craters on Europa have been filled by sub-surface water that breaks through fractures in the hard, water-ice crust and then freezes over. Further evidence that Europa's ice crust sits on an underlying ocean comes from iceberg-like blocks seen on its surface. The blocks appear to have separated, rotated, and drifted; somewhat like icefloes on Earth, the structures resemble the break-up of frozen seas on the Arctic. Readings of the moon's magnetic properties made by the Galileo orbiting mission also point to electric currents being produced beneath the icy crust, as would be expected for an electrically conducting, salty, ocean.

How is enough heat generated inside a small and very distant moon to keep a sub-surface ocean from completely freezing over? The answer is the action of gravitational tidal forces between Jupiter, Europa, and Jupiter's other moons. As Europa orbits Jupiter, the gravitational pull between its two hemispheres varies enough to alter slightly the shape of the moon. The surface of Europa is distorted by tens of yards. This motion generates internal friction, which in turn creates heat. Essentially, the interior of Europa is being powerfully squeezed and released, creating heat in the process. The tidal heat is barely enough to melt the ice from the interior, while upper surface layers remain as a rock-hard crust of ice up to 10km (6 miles) thick, with a temperature of -145°C (-230°F). Some estimates suggest that the water and ice-slush ocean below the ice may up to 100km (60 miles) deep. It is also conceivable

The southern pole of Saturn's moon Tethys is viewed in this image from the Cassini spacecraft, from a distance of about 1500km (930 miles) above the surface. Tethys' low density suggests that it is mostly composed of water-ice, with a heavily cratered surface that includes cracks and faults. Several of the craters have exceptionally bright floors, which may be the results of impacts penetrating through the surface revealing layers of fresher water-ice material. Tethys has a diameter of about 1060km (660 miles).

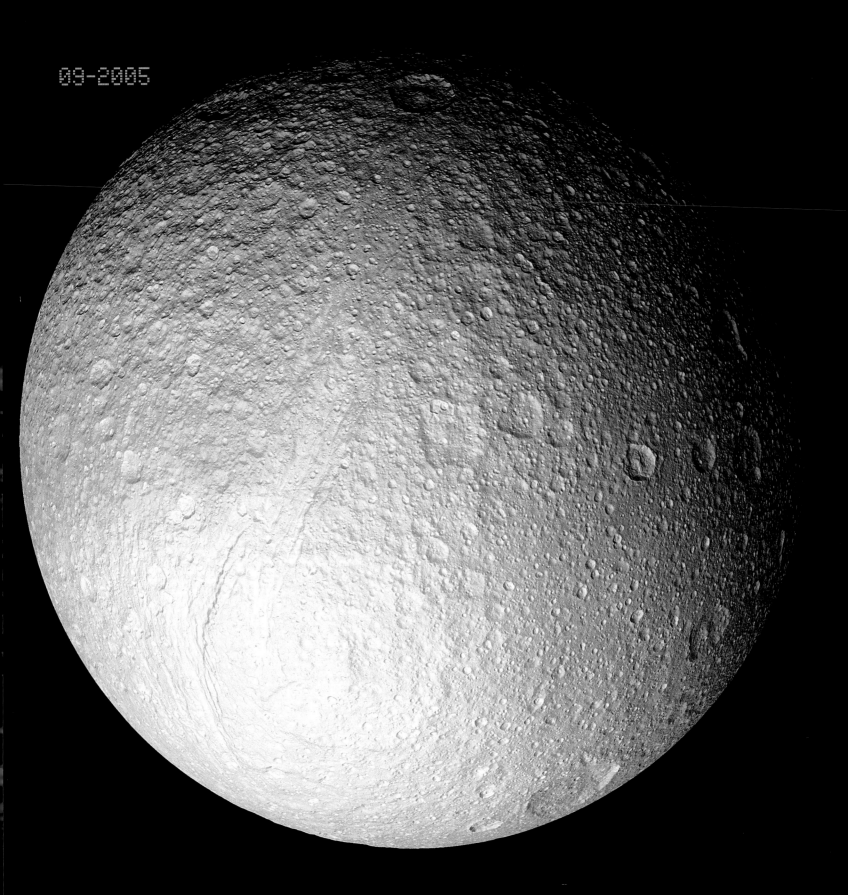

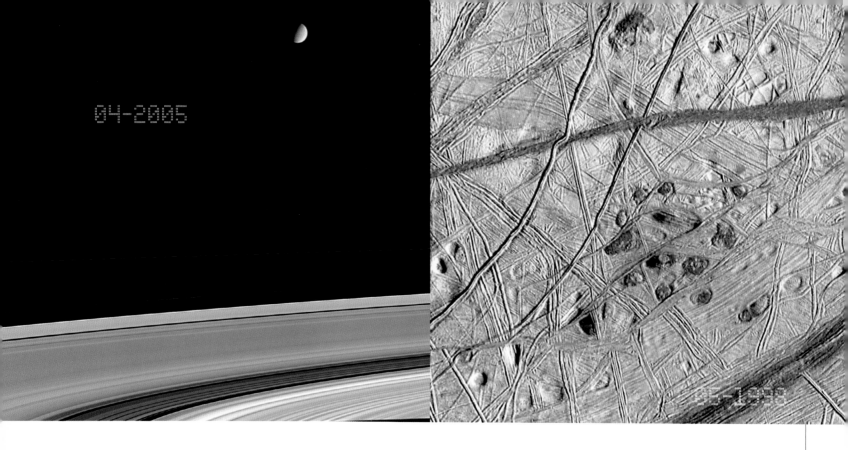

that the ocean floors of Europa have local hot spots in the form of geothermal vents, similar to those found on the mid-ocean ridges of Earth. We saw earlier that on Earth there are the smoking vents ("black smokers") around which a range of life is known to flourish (see pages 112–15). Even if the tidal stresses on Europa are not enough to make it geologically active and create hydrothermal vents, microbes could live close to the upper water-ice boundary. There is currently no evidence whatsoever as to whether microbial life and bacteria exist in Europa's oceans. However, this icy orb, and others like it, offers an intriguing combination of interior heat, sub-surface liquid water, and perhaps organic compounds. All of which makes it a very significant object for future explorations.

Exploring distant oceans

The evidence from the Galileo spacecraft of vast sub-surface oceans ranks as one of the most significant recent discoveries in the solar system. Further dedicated explorations are now required to advance our understanding of whether these distant moons can sustain life.

One of the missions being planned by NASA is a large spacecraft called the Jupiter Icy Moons Orbiter, or JIMO for short. It will aim to study in close-up the substantial oceans of three of the giant planet's moons. Scheduled for launch sometime around 2015, JIMO will travel more than 700 million km (430 million miles) to Callisto, Gaymede, and then finally to Europa. Powered by a nuclear reactor, this ambitious mission will orbit each moon for an extended period to compare their detailed compositions and histories. Radar imaging will be used to determine the thicknesses of the upper layers and look at the sizes and depths of the sub-surface oceans. The spacecraft will also carry instruments to map regions where organic and biologically important chemicals may reside on the surface. The ultimate goal of JIMO would be to identify potential landing sites for follow-up space missions carrying drilling devices and perhaps even for submarine explorers.

Understanding the remarkable surface features of icy moons in the outer solar system is a prerequisite for the success of lander spacecraft, which would be very expensive to build and fly. A fascinating scenario is presented by the dark reddish blotches imaged on the surface of Europa by the Galileo spacecraft. These blotches are known as lenticulae and resemble 10km (6 mile) freckles on the icy crust.

The origins of the lenticulae remain enigmatic, but one interesting hypothesis proposed by astronomers is that they could be the result of the upwelling of warmer water from the ocean

Saturn's ice-covered moon Enceladus is seen in the image on the left, together with a glimpse of the giant planet's awesome rings. The common link is that both the moon and the rings are almost pure water ice. Over time, however, the rings have suffered greater contamination from free-floating dust particles, which provide the varying colours of the rings. The image on the right shows intricate surface structures on Jupiter's moon Europa. The Galileo spacecraft revealed "freckles", or spots and pits, 10km (6 miles) across, possibly the result of warm water churning upwards and cooler surface water-ice descending. These "freckles" may hold clues to the composition of a sub-surface ocean.

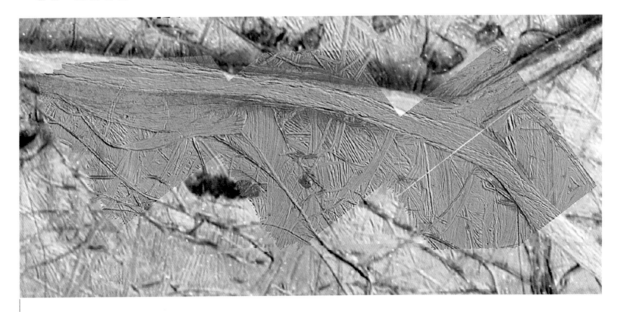

09-1998

This Galileo spacecraft mosaic combines high-resolution colour images and lower-resolution black-and-white views to probe complex features on the icy surface of Jupiter's moon Europa. The bright band stretching from left to right in the image is a feature known as Agneor Linea. This unusual structure is about 1,000km (620 miles) long and 5km (3 miles) wide and its origin is thought to be connected with the substantial ocean of liquid water believed to lie beneath the moon's cracked and frozen crust.

beneath. The dark surface freckles would then mark the tops of narrow "tubes" that bring warmer ocean water up, while colder surface water sinks. Acting like "elevators", the water channels could also transport organisms and nutrients from the lower depths of the oceans. If this explanation for the lenticulae is correct, they would represent superb sites for placing probes on Europa, because the surface ice will be much thinner, making it considerably easier to melt through to the ocean. Submarine-like landers may even be able to ride the vertical currents below the lenticulae to study deep ocean organisms.

A more realistic option over the next decade or so may be to explore the oceans of Europa by listening to its surface ice cracking. Scientists studying the Arctic ocean on Earth have used sensitive microphones to record the noises from ice sheets that vibrate under stress. Large fractures of the ice sheets can generate sound waves that travel for hundred of kilometres

through the ocean beneath. Researchers now believe that these acoustic techniques could be employed to determine the depth of the oceans below Europa's surface layers.

Images such as those returned by the Galileo spacecraft indicate that Europa's crustal ice is subject to numerous and frequent fractures, perhaps even several times a day. The plan, based on Earth ocean science, is to deploy an array of microphones on the surface of Europa to listen for the natural sounds made by the shifting ice. The vibrations created by the ice fractures would be detected acoustically and analysed to provide information on the thickness of the surface layer. Furthermore, detection of echoes would determine the depth of a potential ocean on Europa.

Certainly, if life is present in water lying under Europa's crust, it will be very difficult to access. Any searches would have to be ambitious and pioneering, demanding the latest scientific innovations. These challenging pursuits are, however, likely to generate exciting technological spin-offs that would have beneficial impacts on other aspects of our lives on Earth.

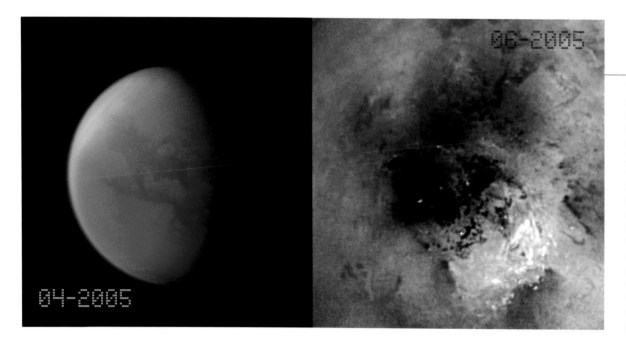

Two pioneering views of Saturn's remarkable moon Titan from the Cassini-Huygens mission. On the far left is a false-colour composite image taken in the infrared waveband, to probe through the Moon's dense atmosphere of mostly nitrogen, oxygen, and methane. The green areas are regions where the spacecraft is able to see down to the moon's surface. The image on the right shows an intriguing dark structure towards Titan's southern polar region. Though its precise nature remains enigmatic, its shore-like boundary may be evidence for an ancient, or even present, liquid hydrocarbon lake.

Cassini Mission at Exotic Titan

For some time now, astronomers have speculated that one of the most promising candidates for habitability beyond Earth is Saturn's spectacular moon Titan. Larger than Mercury, and the second-largest moon in the solar system, Titan is the only planetary satellite with a substantial atmosphere. Indeed, its atmosphere is similar to, though thicker than, that Earth's, with up to 98 percent nitrogen and many compounds that provide the basis for organic chemicals. Despite extreme surface temperatures of -180°C (-290°F), Titan is a smoggy world that holds fascinating prospects for some forms of biology.

By a complex sequence of chemical processes, hydrocarbons such as methane, ethane, and propane are present in Titan's atmosphere. The photochemical processes at work are thought to be similar to those that occurred on Earth almost four billion years ago. Titan today has preserved this primordial composition and is widely regarded as a remarkable natural laboratory in which to study the origin of life. The hope is that probing investigations of Titan will provide fresh clues on how life developed on Earth. Titan's surface is completely obscured from direct view

by its dense atmosphere, which exhibits a reddish haze as a result of the chemical smog that forms when ultraviolet light from the Sun breaks up the molecules of methane. The broken-down products then form other complex compounds, especially ethane. Eventually these photochemically produced compounds overload the atmosphere and start to filter down toward the surface as a non-stop drizzle of toxic rain, which would probably form exotic lakes of ethane and methane.

The fascinating possibilities of a smoggy atmosphere and organically rich seas motivated a mission to Saturn called Cassini-Huygens, which would also closely study Titan. Launched in October 1997, the two-ton spacecraft began orbiting Saturn in July 2004 and flew as close as 1200km (3940 miles) to Titan shortly afterwards. During its close approach, the Cassini orbiter used an array of instruments to study Titan's atmosphere, including radar and infrared imaging to penetrate the smog and map the surface. In addition, on Christmas Day 2004, a small vehicle carried by Cassini called the Huygens probe detached from the orbiter and, after a 21-day descent, landed (aided by a parachute) on Titan on 14 January 2005. The Huygens probe beamed back pioneering images and measurements during its descent through Titan's atmosphere, and had sufficient battery power on landing to continue transmitting data for nearly an hour. The probe is now the most distant craft ever landed on a planet or moon. The Cassini orbiter performed a further three flybys of Titan up to April 2005.

So what has been learned so far from the Cassini-Huygens mission? The orbiter itself peered through the hazy atmosphere to reveal surface details of Titan never before seen. The intervening haze turned out to be highly structured, with up to 12 separate layers that extend more than 500km (300 miles) above the moon's surface. The atmospheric constituents included complex chains of hydrocarbons, which combine to produce an orange hue in the lower atmosphere. The largest collection of clouds witnessed by Cassini gathered around the southern pole region of Titan and revealed evidence of meteorological activity. The clouds appeared to change in time with patterns that indicate cloud formation, evaporation, and rain. In short, this is a sort of methane cycle akin to the Earth's water cycle. Dark patches on the surface provided evidence of a hydrocarbon lake, possibly being fed by more active white methane clouds.

Radar imaging was used to penetrate the dense Titan atmosphere, reaching down to the surface. Narrow and elongated features were revealed that may be interpreted as the channels carved by rubble as it is "washed" down sloped terrain by the flow of liquid methane. The rough rubble appears bright in the radar images. Some of these features extend for hundreds of miles and may possibly be transient rivers. One particular radar bright (therefore generally rough) land area on Titan is the Xanadu Region, which is thousands of miles across. Contrasting dark spots are also superimposed over this region, which are thought to be due to very smooth, probably liquid, regions, or lakes.

During a parachute descent of more than two hours, the Huygens probe recorded considerable turbulence and west-to-east wind speeds of up to 450km/h (280mph). This "super-rotation" means that the atmosphere of Titan is moving faster than its surface. The probe transmitted in-flight images of small channels, ice bedrock, and organic material washed into valleys by methane rain. Measurements made on landing revealed that the probe came to rest on a relatively flat surface with a texture somewhat like wet clay or loosely packed snow. Scientists suspect that the craft impacted onto soil made of water ice held together by a sticky hydrocarbon such as methane. High-resolution images revealed highland regions that had an intricate network of drainage channels, probably carved by liquid methane, that ran into flat lowland regions, such as river beds. An image of a lowland region scattered with 50 small ice-stones up to 15cm (6in) in diameter was taken after the probe landed. The fact that there are no larger rocks suggests that only small pebbles are easily moved from the surface.

During its brief life on the surface of Titan, the Huygens probe additionally captured images of smooth ice blocks that must have been eroded into their rounded forms by wind or the flow of liquid. Overall, the new observations emphasize the role of methane in shaping the surface of Titan, creating ponds, elongated islands, and patterns resembling short coastlines. The remarkable moon is being shaped by complex geological processes that combine stormy conditions in a rich and dynamic atmosphere with a fascinatingly varied surface. The current consensus from the Cassini-Huygens mission is that while a large scale hydrocarbon ocean is unlikely to be present on Titan, it does host liquid methane in small lakes and rivers on its surface. However, with only a tiny percentage of the moon's surface mapped so far, there is no doubt that further exotic features await discovery. The Cassini mission will now continue to monitor Titan, generating vast arrays of data that will take scientists many years to analyse and interpret.

We noted at the beginning of this chapter that liquid water offered the best outlook for life as we understand it. The discovery of liquid methane and ethane on Titan is exciting, but we need to acknowledge that at freezing temperatures the chemical reactions in these media would proceed at a very slow rate. Other compounds that are part of the chemistry for supporting life would also be more difficult to dissolve in hydrocarbon lakes than in liquid water ones. It seems therefore that if microbial life is present on Titan, its rate of metabolism would be unrecognizably slow. There is no doubt, however, that the combination of organic molecules in Titan's atmosphere, ultraviolet energy from the Sun, and ethane-methane lakes, add up to a fascinating environment that over billions of years could potentially have "cooked up" some form of biology. Most profound, perhaps, is that if any life exists in the truly alien environment of Saturn's moon Titan, it must be unlike anything we know on Earth.

Recognizing the Signs of Life

We have gradually pushed the frontiers for potential life habitats from our neighbour Mars, to the ocean-bearing moons of Jupiter, and on to Saturn's moon Titan. Life in these exotic worlds could be very different from what's found on Earth, so the challenge is understanding what to look for and recognizing the essential biosignatures.

The basis of astrobiology is that we appreciate and understand the significance of an object or substance elsewhere in the solar system whose origin implies a biological process. Such a biosignature would reveal past and present life, either via in situ measurements, samples returned to Earth, or a study of atmospheres. On Earth, fossils in sedimentary rocks provide a record of past life, and it may be that similar finds await us in the dried-up lakes and streams of Mars. Other biosignatures may be complex organic molecules that can be formed only in the presence of life. Life on Earth is based on carbon and nitrogen, the chemical breakdown of which releases methane and ammonia. Bacteria on Earth derive energy from carbon dioxide and hydrogen to make methane. The search for these gases on Mars is a path toward discovering bacterial life there.

During 2005, scientists using the Mars Express spacecraft confirmed earlier detections of methane in Mars' atmosphere made using the Keck telescopes in Hawaii. Methane is not a stable or long-lived gas in the Martian atmosphere and must be replen-ished. We are almost certain that the gas is being generated today. A complication is that volcanic activity may be a source of methane produced by sub-surface lava. An exciting alternative view is that microbes, such as methanogens, are producing the methane. Current missions, including the Mars Exploration Rovers, are not equipped to resolve the issue of whether the origin of Mars' methane is due to geological or biological activity.

Despite years of hunting, the equally telling gas ammonia has not yet been confirmed on Mars. Its presence in Mars' atmosphere would indicate the bacterial breakdown of proteins, because ammonia is not easily produced in geological processes such as volcanoes. The detection of biosignatures on Mars requires more sophisticated equipment to be placed on the planet. An important biosignature on Earth is the reservoir of oxygen in its atmosphere. The discovery of an abundance of oxygen in another world would raise the spectre of life, particularly if it were present in greater quantities than other atmospheric gases. It seems that life on Earth has been present for billions of years. The lessons from Earth, Mars, and the icy moons of Jupiter, are that the ideal biosignature is one that is not only most probably created by life, but is also unlikely to be produced by nonbiological processes. We need to assemble an extensive database on how to recognize, measure, and interpret biosignatures that can reveal life in diverse habitable conditions.

Water ice can be found in some of the most unexpected regions of the solar system. The red patches in this radar-based image from the Goldstone Antenna, USA, show highly reflective material near the north and south poles of Mercury. It seems that even this Sun-scorched planet has permanently shadowed craters that remain cold enough to accumulate substantial amounts of water ice, probably originating from crashing comets.

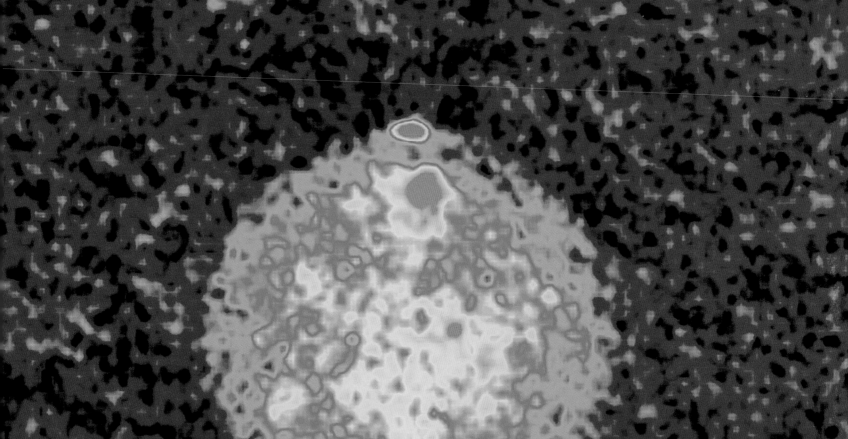

In 2004, Cassini became the first spacecraft to enter into an orbit around Saturn. From this vantage position, the sophisticated observatory has been securing astounding new images of the planet's majestic system of rings. In this unusual view, the shadows cast by the rings are seen projected onto the planet's northern hemisphere.

Rings Around the Solar System

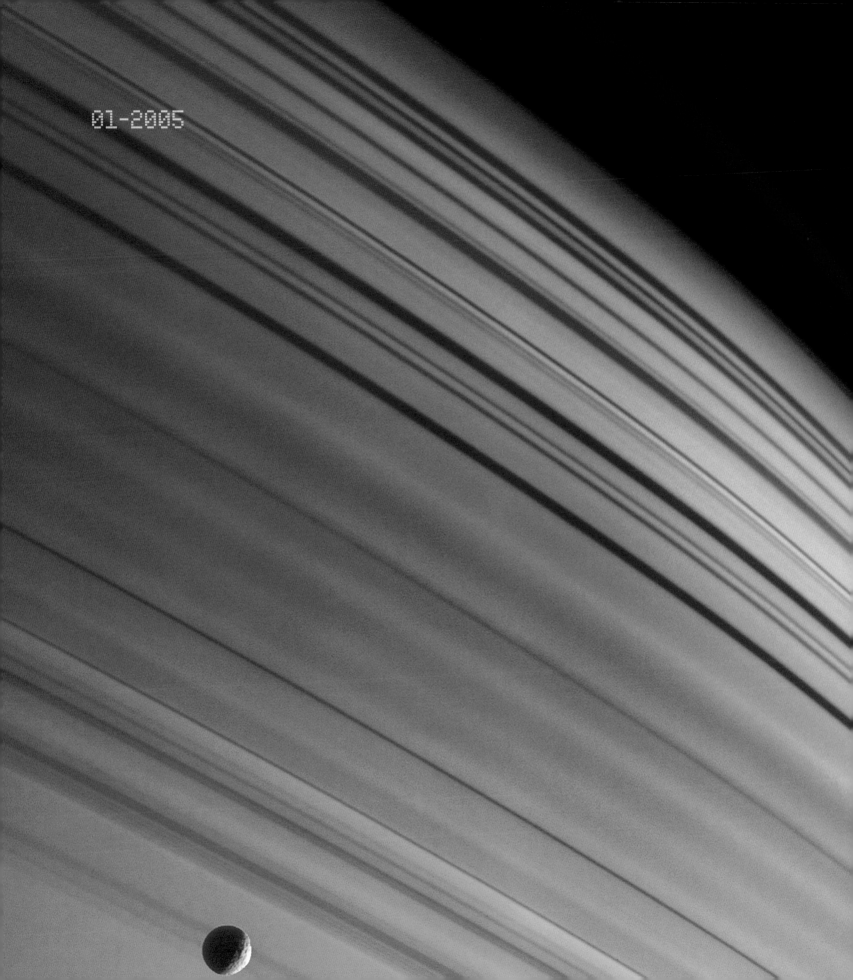

Among the grandest structures in the solar system are the wondrous rings that girdle each of the four giant gas planets. From the brilliance of Saturn's ring system to the darker, more tenuous structures around Jupiter, Uranus, and Neptune, these features provide fascinating insights into the origins of the solar system. Detailed studies of the rings reveal important information about the material from which the planets themselves were formed and evolved billions of years ago. The main focus of this chapter is on the magnificent new images of Saturn's vast ring system acquired since 2004 by the Cassini spacecraft mission. These latest observations provide remarkable new clues to aid our understanding of the structure and composition of the rings, and the dramatic influence of the small moons embedded within the disc of icy particles and rocks. The fresh views of Saturn can be compared with the latest imagery of the rings of Jupiter, Uranus, and Neptune from the Hubble Space Telescope and giant telescopes on Earth. The planetary ring systems may be a transient phenomenon, perhaps lasting only several hundred million years before they dissipate, or are dragged into the host planets. We may truly be fortunate, then, to be present at a time when our solar system is so richly decorated with these stunning spectacles. On an even grander scale, we also take a look here at the vast belts of rocky bodies that form doughnut-shaped structures beyond the orbits of Mars and Pluto.

A serene true-colour view from the Cassini spacecraft of Saturn's atmosphere, where blue light is scattered by gases in the uppermost layers of the planet. The long dark lines superimposed on the blue backdrop are shadows cast by Saturn's intricate rings. The beautiful scene is completed by one of Saturn's moons, Mimas, drifting in at the bottom of the image.

Lord of the Space Rings

Even through a good pair of binoculars or a small telescope, Saturn presents a breathtaking view that is noticeably different from the other planets in the solar system. Its majestic rings have been a source of wonder ever since Galileo Galilei first saw them in 1610 through the earliest telescopes. The optical quality of his 17th-century telescope was not good enough, however, to clearly identify the image or reveal the planet's ring system. The great Italian astronomer, like several other scientists after him, was confounded by what he saw and uncertain about what he was viewing. In a coded anagram, Galileo recorded that he had "observed the highest planet tri-form", thinking that Saturn was flanked by two smaller planets. Six years later he described the "appendages" as two great handles that extended toward the giant planet. Gradually, telescope optics improved and the planet came into sharper view. In 1659 the Dutch astronomer Christian Huygens realized the "handles" described by Galileo were a thin, flat disc that encircled the planet. By 1675 Jean-Dominique Cassini had discovered the first major structural feature, about two-thirds of the way out from the inner edge of the disk. Viewed from Earth the structure looks like a gap in Saturn's rings, and it is named the Cassini Division (after the astronomer). We know today that there is actually dark ring material in this "gap".

For the next couple of centuries, scientists and mathematicians were perplexed by what the rings were made of. By the middle of the 19th century, and following earlier suggestions by a scientist named Pierre Laplace, the physicist James Clerk Maxwell presented conclusive arguments that Saturn's rings were neither solid, liquid, nor gas. He stated that the rings are composed of countless millions of small particles, all orbiting independently, in much the same way as the way the Moon orbits the Earth. This theory was confirmed in 1895 when measurements indicated that the inner edge of Saturn's rings was spinning faster than its outer edge. This difference in orbital speeds would be impossible if the rings were a single, solid, plate-like object.

Even now, however, extensive telescope observations have still not resolved mysteries concerning the origin and complex structure of the rings. The first spacecraft to fly near Saturn was Pioneer 11 in 1979, followed by the Voyager 1 and 2 missions, which made their closest approaches in 1980 and 1981 respectively. Significantly, in 2004 Cassini became the first spacecraft to go into orbit around Saturn. Thus began a four-year mission to study the planet and, of course, the stunning rings.

Alphabet rings

Saturn's rings span an enormous distance. The total diameter of its more tenuous outermost ring is nearly 1 million km (620,000 miles); that is equivalent to almost two and a half times the

These Hubble Space Telescope images acquired over four years, show differing views presented by Saturn's rings as the planet's orientation about its axis changes. In a similar manner to the Earth's, Saturn's equator is inclined relative to its orbit by about 27 degrees. As the planet orbits the Sun, the rings are presented to us from almost edge-on and very thin, to very nearly face-on as winter approaches its northern hemisphere.

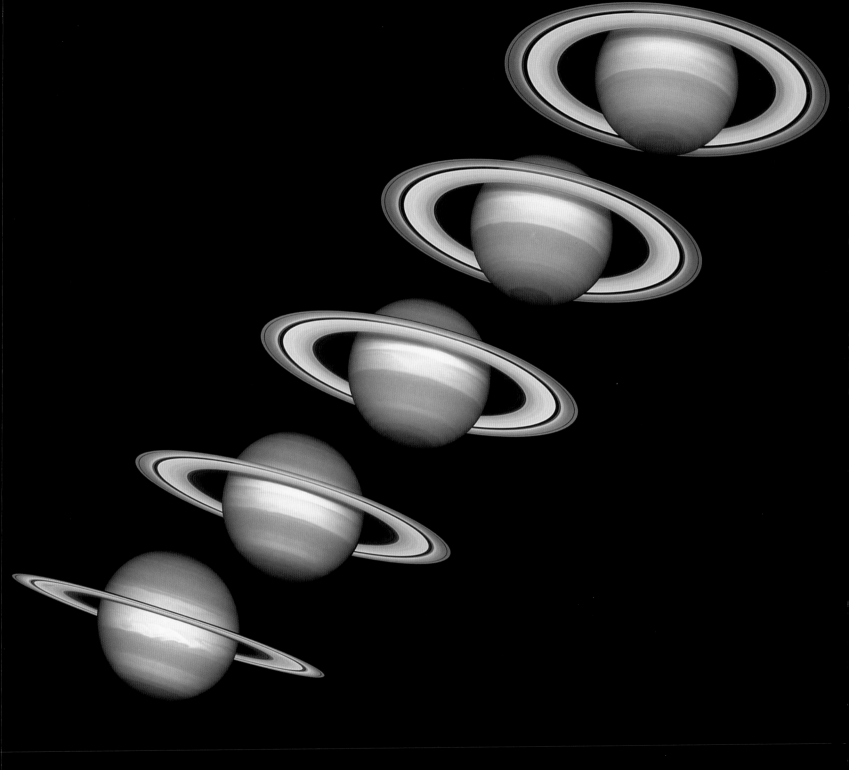

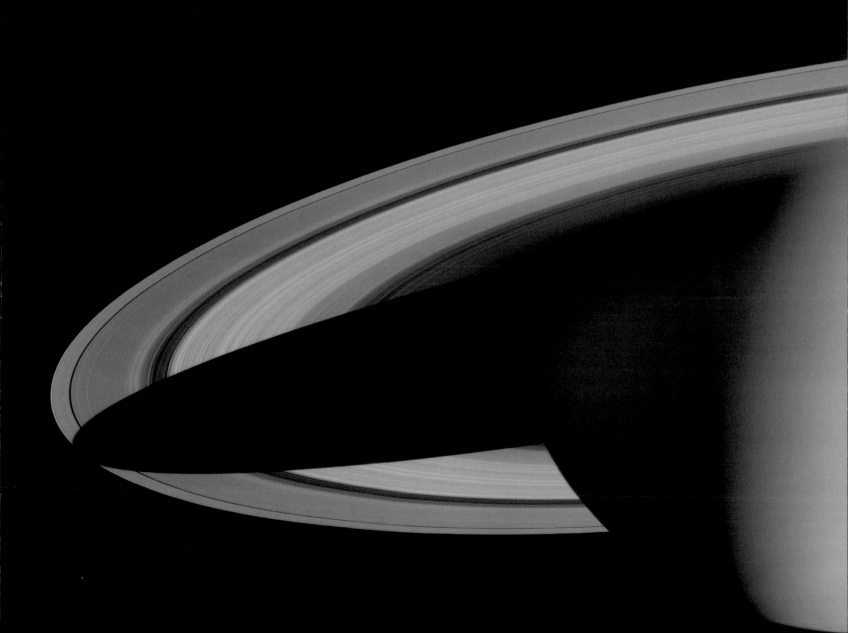
10-2004

distance from the Earth to the Moon. Surprisingly for such a wide structure, most of its rings are only a few tens of yards thick. If a model of the Saturn system scaled the diameter of its major rings to 50m (164ft), the corresponding ring thickness would be about that of a sheet of paper. The rings consist primarily of ice particles ranging in size from microscopic dust to tumbling house-sized boulders. There are probably very few kilometre-sized constituents. The icy material is highly reflective, which is why Saturn's rings are brighter and far more spectacular than those of the other giant gas planets.

Saturn's ring system is made up of seven major divisions that have been given alphabetical names in order of their discovery. The Voyager spacecraft revealed that each of these primary divisions is in turn composed of thousands of individual narrow ringlets. Moving outward from the planet, the main broad and visible ring structures are designated C, B, and A. The "gap" discovered by Jean-Dominique Cassini separates rings B and A. The inner ring, C, is about 17,500km (10,800 miles) wide and made of numerous dark ringlets, although the ring itself is fairly transparent. B is the inner of the two brightest rings, and made up of ringlets barely 15km (9 miles) wide. It is one of the most opaque of Saturn's major ring structures, and is more than

25,000km (15,500 miles) wide. A peculiarity of ring B is the presence of dark temporary spoke-shaped features, known as drifters, that move in strange paths. These drifters are 20,000km (12,400 miles) long and are thought to be very fine dust particles suspended above the ring. The dust probably originates from meteorites that have punctured the rings, and is lifted by electrostatic forces.

The 14,600km (9070 mile) wide ring A is the outer of Saturn's two brightest rings. Variations in brightness and colour are caused by the combined effects of differences in particle sizes and the way they scatter sunlight. The water ice in these ring particles is also contaminated to different degrees by rock, dust, and other carbon-based materials. The high-resolution images from Cassini have reveal pronounced colour variations that will allow scientists to understand the precise composition of different sectors of the ring system. There is a small gap, called the Encke Gap, near the outer edge of ring A, inside which resides Saturn's tiny moon, Pan. The gravitational action of Pan is thought to keep the gap clear of particles.

Ring F is a narrow feature – less than 500km (310 miles) wide – just outside A. Discovered by the Pioneer 11 spacecraft in 1979, it is probably the strangest of Saturn's main rings,

Two images from the Cassini mission. On the left is a false-colour ultraviolet image of the outer region of Saturn's C ring and the inner portion of the B ring. The colour-coding emphasizes the varying compositions of the rings. Particles more contaminated with dust are indicated in red, and cleaner ice particles are shown in turquoise. Radio signals were used to construct the false-colour image on the right and to obtain information about the sizes of particles in Saturn's rings. In the red regions most particles are more than 5cm (2in) across, while the green and blue shades highlight progressively smaller particles – about 5cm to 1cm (2in to less than 0.5in) across.

Previous page
This spectacular full-span view from Cassini is one of the most detailed images ever assembled of Saturn and its vast rings. Over 100 high-resolution images have been combined to capture the full 1 million km (630,000 miles) span of the icy rings, which is equivalent to two-and-a-half times the distance from the Earth to the Moon.

because of its intricate structure. The two outermost rings, G and E, are thicker and fainter than the inner rings. G was discovered in 1980 by Voyager 1, while the more distant presence of ring E was also confirmed by Pioneer 11. The 8,000km (4,970 mile) wide ring G is extremely tenuous, likely to be made up of tiny particles. E is Saturn's most distant, and made of even more microscopic particles, which are blue in colour. Several of Saturn's moons reside within this ring and play a significant role in organizing and supplying the ring material. E is the thickest and most extended of the rings, with a width of around 300,000km (186,400 miles).

Cassini at Saturn

In June 2004, the Cassini spacecraft arrived at Saturn and began to orbit the giant gas planet, heralding an exciting four-year mission of discovery. Named after the astronomer Jean-Dominique Cassini, the mission represents one of the most sophisticated ever launched to a planet. In addition to the orbiter, it also carried the Huygens probe, whose descent to Saturn's largest moon Titan was described on pages 132–3. The orbiter itself is packed with a formidable suite of instruments designed to investigate all the major aspects of the Saturn system. High-definition pictures are obtained using long-and short-focal-length cameras, which can detect light beyond the visible ranges of the human eye. Spectrometers are used to decode the chemical composition of the planet, its rings, and its moons. Additional instruments are designed to make in-situ measurements of Saturn's magnetic and electrical properties.

The Cassini orbiter and the Huygens probe had a combined weight of 5,712kg (12,590lb), which is massive by spacecraft standards. In order to propel it into the outer solar system, the mission scientists used a technique known as "gravity assist". Following its launch in 1997 on an enormous Titan IVB rocket, the spacecraft passed by Venus twice, followed by the Earth in 1997, and finally made a pass by Jupiter in 2000. During each of these close planet approaches, Cassini received a "sling shot" boost from the gravity of the planet to increase its momentum and accelerate it toward Saturn.

The Cassini spacecraft conducted its first sequence of observations of Saturn's rings during 2004, returning the most detailed images ever seen of their intricate architecture. Over its lifetime, Cassini will orbit the planet numerous times and return thousands of images. It has also started to conduct pioneering radio

and infrared scans to provide new information about the thermal properties and composition of the rings. An early intriguing discovery from the spacecraft concerned the ingredients of the Cassini "gap" between the A and B rings. Most of Saturn's rings are composed of relatively bright water ice, but the latest findings show that the Cassini division itself is dark not because there is nothing there, but because it contains more dirt than ice. Furthermore, the dark particles are of a similar make-up to that of the material found on Saturn's moon Phoebe. This information gives renewed backing to the notion that the Saturian rings might be leftovers of an ancient, shattered moon.

Another surprise from Cassini was the detection of large amounts of oxygen at the ring edges. It seems that the spectacular ring system may have its own tenuous atmosphere, totally separate from the planet itself. By swooping close to the rings, the spacecraft has measured that the atmosphere above the rings is mainly made up of oxygen molecules. A likely origin is that some of the water in the rings is broken up, and the light hydrogen gas escapes, while the heavier oxygen is held back by the gravity of the ring system.

One of the fundamental questions that scientists are trying to answer regards the dynamic relationship between the austerely beautiful rings and the numerous tiny moons that nestle in the great disc of icy material. Acting like shepherds, the moons gravitationally corral the ring material, forging narrow clearances and complex structures. Together, the forces between the icy particles and the "shepherd moons" can herd the rings into kinked and spiralling ringlets. Cassini observations secured during 2005 show how Saturn's miniature satellites Prometheus and Pandora are influencing the enigmatic F ring. The two moons squeeze ring particles together and carve out curious braided- strand features, which seem to join and separate out around the circumference of ring F. The outer of the two moons, Pandora, is 85km (53 miles) wide and prevents ring F from speeding away from Saturn. The inner moon, Prometheus, is 100km (62 miles) across and acts in a counter direction to stop the ring being pulled toward the gas planet. The images from Cassini have thrown up the further surprise that ringlets flank the ring F, forming a spiral pattern, thought to be the consequence of minuscule moons that cross ring F. The spiralling ringlets, and Cassini spacecraft observations of sports-field sized clumps embedded within rings, are now thought to represent the fingerprints of unseen moons in Saturn's rings. The sculpting

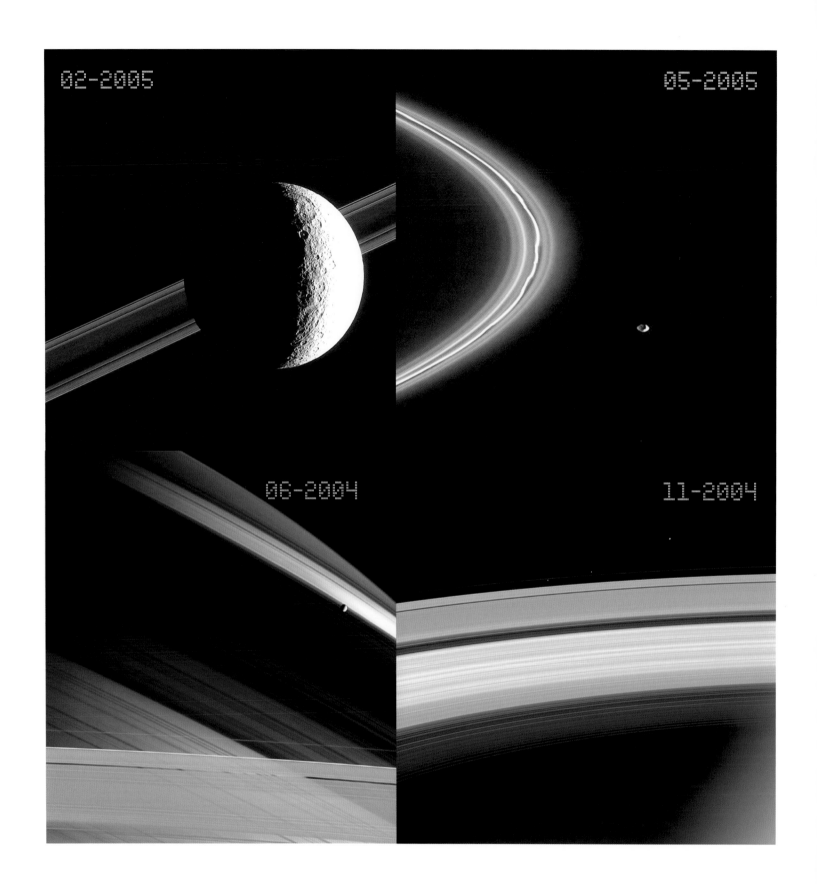

Opposite

Top right: The influence of Saturn's numerous moons in shaping the planet's ring structure is highlighted in this set of four images from the Cassini mission.

Top left: Sunlit Rhea is seen against the dense ring B. The tiny, 84km (52 mile) wide Pandora shepherds dust-sized particles into the narrow ringlets that border Saturn's ring F.

Bottom left: The gravitational influence of Mimas is responsible for clearing material in the Cassini Division between rings A and B.

Bottom right: Prometheus, Pandora, and Janus, just visible in the upper half of the image, join forces to control and structure Saturn's ring F.

is on a smaller scale than anything that has been imaged by previous spacecraft. The embedded moons may be just 5km (3 miles) or less in diameter.

Detectors on board Cassini also listen out for radio waves that are helping scientists to further decipher the structure of the system of rings. The radio data highlight the tremendous variety of particles that reside in the rings, and the wavy forms they adopt under the gravitational influence of nearby moons. An unexpected discovery was the tone of radio sound waves emitted as the spacecraft passed over Saturn's rings. The tones lasted for a few seconds at most and sounded like musical notes when their frequencies were reduced to bring them into the range of human ears. The radio "tunes" are thought to be created by energy released when 1cm (0.4in) objects, or meteoroids, crash into the ring material. This is the first time that the constant bombardment of meteoroids into Saturn's rings has been directly detected.

During late 2004 Cassini also secured the most remarkable pictures ever obtained of Saturn's rings, using ultraviolet light. The ultraviolet imagers were able to "see" ring structures as small 97km (60 miles) wide, improving by a factor of 10 on what Voyager 2 had previously obtained in this waveband. As a clue to the origin and evolution of the rings, the ultraviolet images show that there is more ice in the outermost regions of the rings, while the inner, sparser ringlets are more contaminated by dust particles.

Origins of the rings

The many thousands of narrow ringlets that make up Saturn's ring system clearly form a very complex but amazingly stable structure. We have seen that the icy ring constituents range from dust-sized particles to boulders several yards across. The action of gravity between the ring particles, and the rings and Saturn's moons, has carved complex structures such as narrow clearings, wave-like motions, and clumps. How did such a phenomenal structure form around a planet?

Before the era of spacecraft missions to Saturn, some scientists believed that the rings may have formed at the same time as the planets. Instead of coalescing into the main body of Saturn, remnants of material left over billions of years ago when the solar system was formed, they thought, stabilized into a ring encircling the planet. Over the past few years, however, this theory has appeared to be flawed. There is now a growing real-

ization that Saturn's rings cannot be billions of years old. Quite simply, the fact that they are so bright tells us that the rings must be young, perhaps only a few hundred million years old. It may be the case that Saturn did not have rings at a time when dinosaurs dominated the Earth. The theory is that as Saturn orbits the Sun, the rings gradually accumulate the dust that prevails in the solar system. The older the rings, the greater the dust content and thus the darker their composition. Therefore if the rings really were billions of years old, they would not have the brilliance they display today.

The relative youth of Saturn's rings has led to an alternative theory that these great structures are the remains of a moon torn apart under the influence of gravity. The notion is that a small, perhaps 250km (150 mile) wide, moon approached very close to Saturn and the giant planet's immense gravity pulled harder on one side of the body than the other. This difference was enough to set up a gravitational tide that shattered the moon. The resulting debris then settled into concentric orbits, eventually forming a set of rings. This scenario of a catastrophic origin is not problem free, however, and the many details of how the rings would have formed have still not been established. It is also possible that the destruction of the moon may have been caused by another violent event, such as a collision with another satellite, or perhaps even a large comet. Observations such as those obtained by the Cassini mission also suggest that the material in the rings may even be continually replenished by chunks ejected from Saturn's moons following the bombardment of meteorites.

These ideas for the dramatic creation of Saturn's rings raise the remarkable prospect of new ring systems developing elsewhere in the solar system sometime in the future. Mars' tiny potato-shaped moon, Phobos, is, for example, being gradually drawn inwards towards the planet by gravitational tides. In about 50 million years from now Phobos will be close enough to Mars that stresses from the planet will destroy it. The result could be a faint new ring system. Potentially even more spectacular is that Neptune's 2700km (1670 mile) moon, Triton, is also slowly spiralling towards its planet. In about 100 million years time, Triton will be ripped apart by Neptune's immense gravity, leading to the creation of a fantastic new ring system. It is even possible that Saturn's rings may have collapsed or dissipated by then, leaving the planet Neptune as the new Lord of the Rings.

Circling the Other Gas Giants

Saturn is not the only ringed planet in our solar system, although it is the only one we have known about for more than three centuries. In March 1977, astronomers were monitoring the behaviour of light from a star as Uranus moved in front of it. The planet's rings caused flickering in the starlight just before the star was blocked from view by the planet, and again as it re-emerged from the other side. Subsequently the Voyager 1 mission to Jupiter in 1979 and Voyager 2 to Neptune in 1989 revealed that these planets also had rings. In just over a decade it had become clear that rings were a common feature of our solar system. Indeed, they are a key characteristic that must explained by any theory about the origin of the solar system.

Rings around Jupiter

Saturn still has the distinction of hosting the most magnificent and brightest rings of all the planets. The particles orbiting the other giant gas planets are less extensive, fainter, and darker than

The four major rings of Uranus and several of its moons are shown in this rich infrared view from the Hubble Space Telescope. Most of the moons orbit very close to the planet, at only about one fifth of the distance from the Earth to the Moon. The gravity of the moons confines the ring particles into very narrow orbits around the giant gas planet.

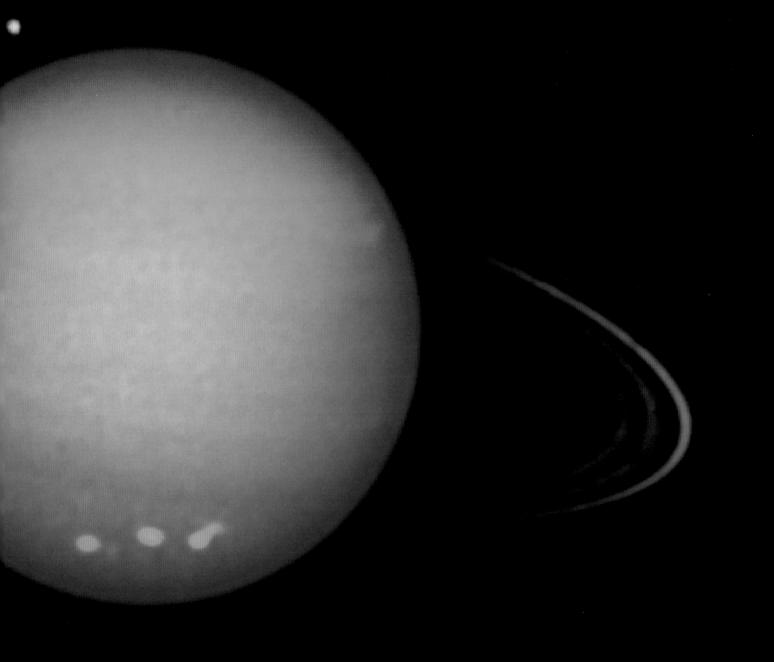

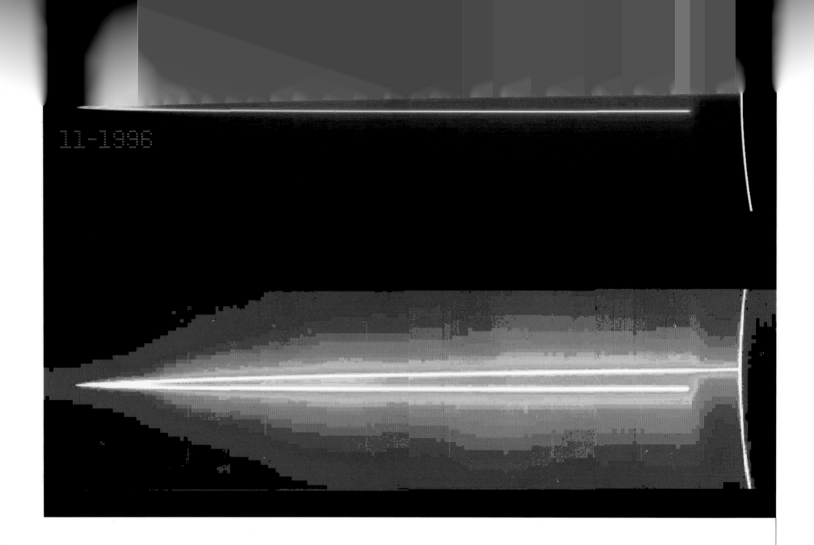

Saturn's. This makes them considerably more difficult to observe from Earth. Following the observations from the Voyager spacecraft missions, however, the ring system of Jupiter is known to comprise three main sections.

The primary ring stretches from 50,000km (31,070 miles) above Jupiter's cloud tops out to a width of 6440km (4000 miles). As is the case with Saturn, this ring is very thin – barely 30km (19 miles) from top to bottom. The orbits of two small moons, Adrastea and Metis, nestle within the main ring and it is likely that they provide a source for the dust that dominates this structure. The innermost region is more of a cloud-like ring than a disk. It forms a doughnut-shaped 10,000km (6215 miles) thick halo extending right down to Jupiter's cloud tops.

Beyond Jupiter's main ring is a pair of structures, which are known as the Gossamer Rings. These are extremely faint circles bounded by the orbits of the moons Amalthea and Thebe. Images from the Galileo spacecraft in the 1990s provided direct evidence that these moons are the source of dust for the

Gossamer Rings. It seems, therfore, that Jupiter's rings are largely made from dusty debris that has been ejected from the numerous small moons that are struck by asteroids or comets as they orbit the giant gas planet.

The narrow rings of Uranus

The observation in 1977 of the irregularities in a star's light as Uranus passed in front of it revealed a total of nine densely packed rings. Despite their great circumference of almost 250,000km (155,340 miles), the rings are astonishingly narrow and barely 10km (6 miles) wide. Uranus' rings are thought to be made of 1m (3ft 3in) boulders, composed of dark material that reflects precious little sunlight. The material is similar to that of several small moons that orbit Uranus, which again makes it likely that the moons are the source of the ring particles. The Voyager 2 spacecraft discovered two more faint and narrow rings together with evidence that, like Saturn's small moons influence the shapes of the structures around the planet. In 2005

A pair of views from the Galileo spacecraft revealing the thin strands of material that circle Jupiter. The upper image was obtained when the spacecraft was in Jupiter's shadow, gazing back toward the Sun. The false-colour image at the bottom is designed to emphasize the faint structures of Jupiter's delicate ring system.

A tantalizing glimpse of Jupiter's tenuous rings is provided by this mosaic of images from the Galileo spacecraft. The rings are seen from a distance of more than 2 million km (1,205,000 miles), from within the shadow region behind Jupiter. Sunlight scatters off the tiny ring particles and from haze in the giant planet's atmosphere.

astronomers used the powerful capabilities of the Hubble Space Telescope and the 10m (33ft) Keck II telescope in Hawaii to uncover two new rings around Uranus.

Observations of the rings have revealed faint dusty structures orbiting well beyond the previously charted system of 11 rings. The innermost of the new rings is about 67,700km (42,070 miles) from the planet's centre, and the outer one is at least 30,000km (18,640 miles) farther out. The outer ring is bluer and is thought to contain smaller dust grains, which are being replenished by a 19km (12 mile) wide satellite, known as Mab. Impacts from meteoroids blast dust off the surface of Mab, which then accumulates as a ring around the planet.

Neptune's ring system

In 1989, the Voyager 2 spacecraft mission to Neptune, confirmed that a ring system orbited this the most distant giant gas planet. The outer faint ring, called the Adams Ring, contains three prominent arcs whose origin remains somewhat enig-

matic. The expectation was that material concentrated in the arcs should quickly spread out into a uniform ring. The fact that it hasn't done so, leaving clumps of material in the Adams Ring, is not well understood. It's quite probable that once again we are witnessing the gravitational interplay between the planet's moons and its rings. In this case, Neptune's moon Galatea may well be acting to confine material into arcs.

Other rings detected around the planet Neptune include the Leverrie Ring, which is about 53,000km (32,930 miles) from the planet's centre. A thousand kilometres (620 miles) closer in towards Neptune lies the broader Galle Ring. Our understanding of Neptune's rings has become even more uncertain recently, with the discovery in 2005 that the whole system is more unstable and variable than previously thought. It is quite possible that some of the components of Neptune's planetary rings may even disappear completely over the next hundred years or so.

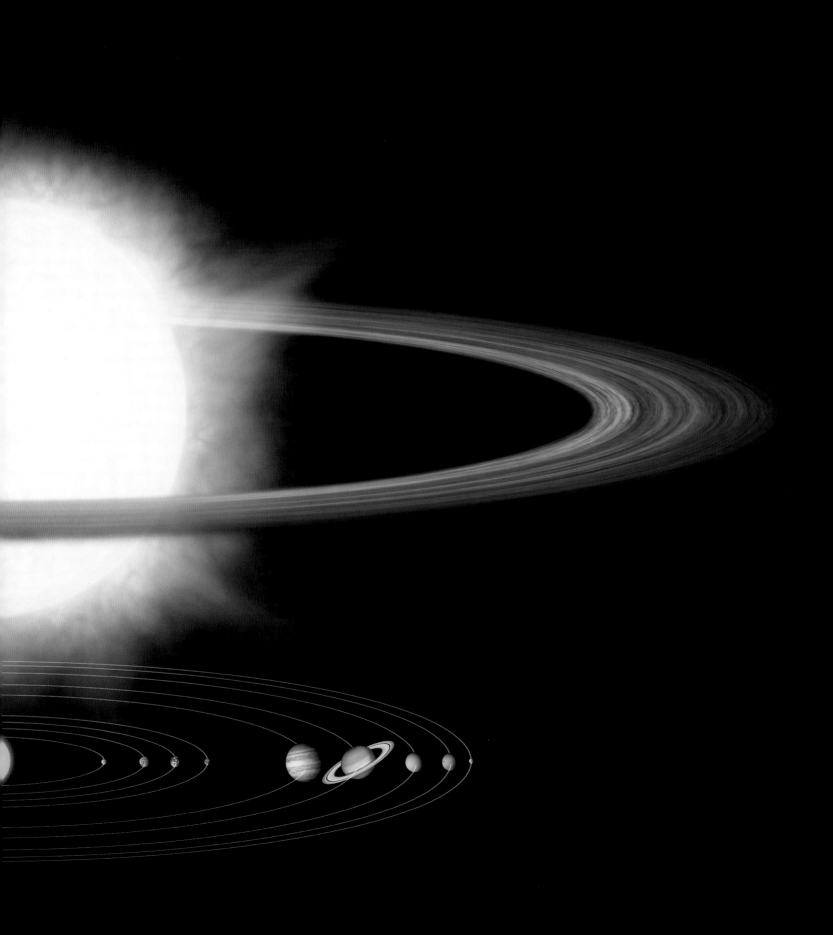

Giant Icy Belts

There are immense circular structures in the solar system that totally dwarf any impressive planetary rings that we have described already. Of considerable recent interest is the Kuiper Belt. This is a disc-shaped region deep in the outer solar system that extends from about the orbit of Pluto to 500 times the distance between the Earth and the Sun. The Kuiper Belt contains numerous icy bodies and represents one of two vast reservoirs of comets in the solar system. (The other is a larger spherical region called the Oort Cloud.) Disturbances such as collisions between the icy objects in the Kuiper Belt can launch them into typically 100-year, highly elliptical or egg-shaped orbits around the Sun. Jets of gas and dust are released as a dusty chunk of ice approaches the Sun, eventually partially vaporizing to form the characteristic tails of a comet.

It is estimated that there may be about 3500 bodies residing in the Kuiper belt with diameters of 100km (62 miles) or greater, as well as countless more smaller ones. Objects in the Kuiper Belt are also occasionally disturbed by gravitational interactions with the giant gas planets, causing them to cross the orbit of Neptune. This places the Kuiper Belt objects in a rather unstable position, from which they could be either ejected from the solar system or thrown further into it. Chiron is an example of such an object currently residing closer than Neptune. With a diameter of 170km (105 miles), if Chiron is ever nudged into an orbit around the Sun, it would warm up and transform into a truly phenomenal comet, considerably more spectacular than comet Halley.

New planets or numerous planetoids?

The study of objects in the Kuiper Belt has gained impetus recently following the discovery of small, icy bodies very similar to Pluto and Neptune's moon Triton. Observations of these remote orbs have raised fresh speculation about the prospects of finding new planets in our solar system, and indeed whether Pluto should be classified as a planet at all.

In late 2002, an object about 1250km (775 miles) in diameter, was discovered in the Kuiper Belt and given the name 2002 LM60 or Quaoar. Shortly after that a body named Sedna was uncovered about 130 billion km (80 billion miles) from the Sun. Although Sedna is smaller than Pluto (by some 600km or 373 miles in diameter), these bodies, or planetoids, have started new discussions among astronomers as to just what is the definition of a planet.

In February 2004, using images taken with the telescopes on the Palomar mountain in Southern California, astronomers uncovered a new Kuiper Belt object named 2004 DW. From its brightness and distance, the object is estimated to be

A revealing image from the Spitzer Space Telescope of the comet Encke with its twin jets (image centre). This comet orbits the Sun every three and one-third years, and as it travels through the solar system between the orbits of Mars and Jupiter it leaves a vast trail of gravel-like debris, seen here as the diagonal band. This material forms a ring around the solar system. Every October the Earth passes through the dust trail, resulting in the Taurid meteor shower.

Previous page
Enormous disks that circle massive hypergiant stars have been uncovered by NASA's Spitzer Space Telescope. Vastly larger than our Sun, these stars reside in a neighbouring galaxy almost 170,000 light years away. This illustration compares the size of these stars and their disks with that of our own solar system. The rings of the hypergiant stars are thought to contain the raw materials for building planets.

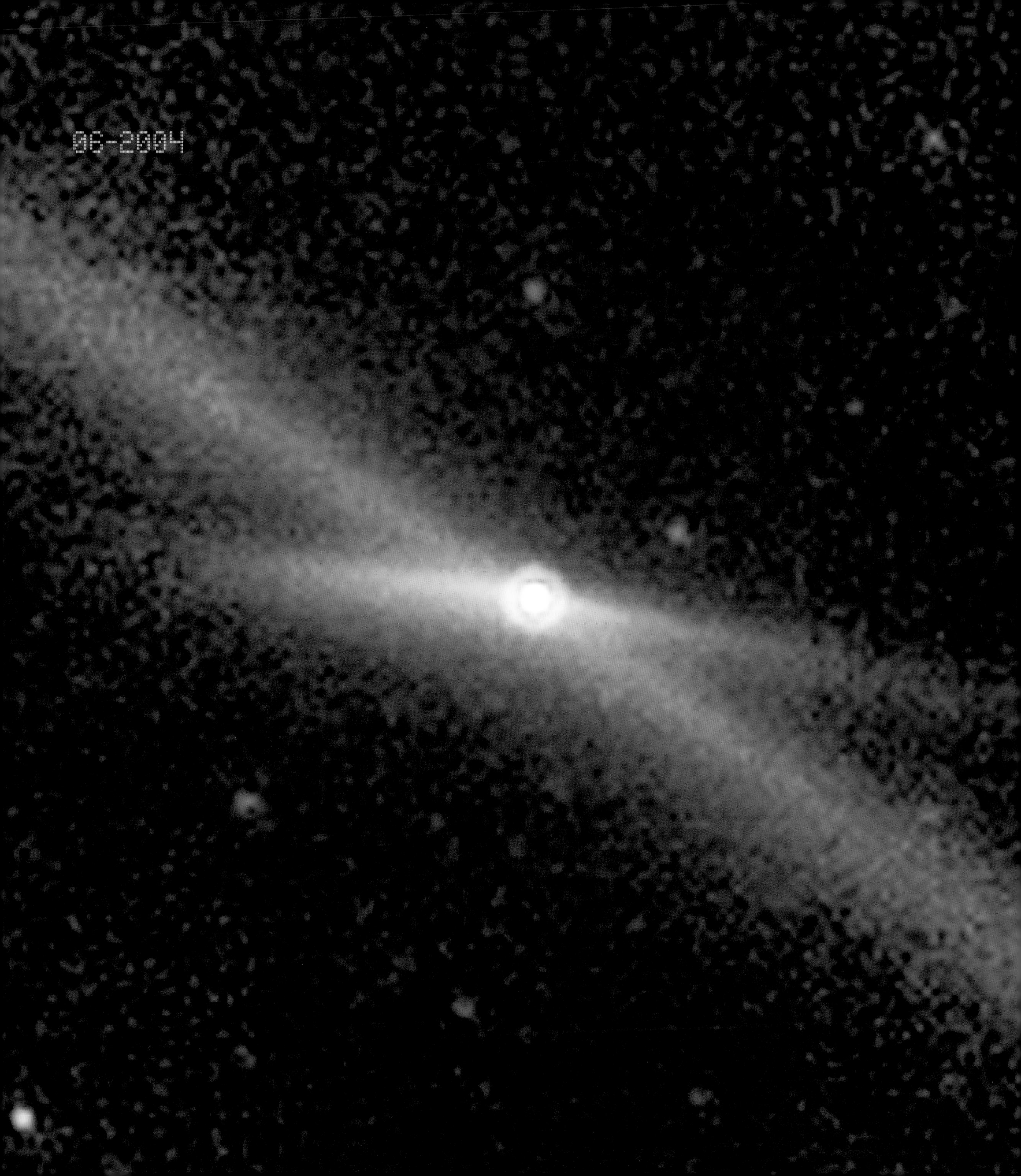
06-2004

1600km (995 miles) across, making it larger than Pluto's moon Charon. 2004 DW is further away than Pluto, some 7 billion km (4.5 billion miles) from Earth. Light travels at 300,000km per second (186,400 miles per second) and takes eight minutes to reach us on Earth from the Sun. In comparison, the Sun's rays would take six hours to travel to the depths of the Kuiper Belt where 2004 DW resides. Like most objects this far out in the solar system, 2004 DW is most probably made of a combination of rock and different types of ice that include water, methane, and carbon dioxide.

One of the most dramatic announcements regarding a Kuiper Belt object came in January 2005, in what was reported across the world as the discovery of a "10th planet". The object in question has the temporary name 2003 UB313. It was found during an ongoing survey of the Kuiper belt, using a telescope in the Palomar Obervatory, California. The telescope is remotely controlled and records numerous images every night. Computers then analyse the images looking for evidence of new objects in the solar system. These objects are identified by their tiny movements against the background of "fixed" stars seen through the telescope.

The reason for the excitement is that 2003 UB313 is the largest body found orbiting the Sun since the discovery of Neptune in 1846. The new object is estimated to be larger than Pluto and is thus also the largest known object in the Kuiper Belt. Its exact size is difficult to determine from its brightness, because we are not sure about the composition of its surface. If 2003 UB313 is made up of a Pluto-like mixture of rock and ice, its diameter could be about 2850km (1770 miles), compared with Pluto's size of around 2220km (1380 miles). Even if the object has a highly reflective surface comparable to fresh snow on Earth, its inferred diameter would still be slightly larger than Pluto's. Because of its substantial size, 2003 UB313 has been hailed as a "new planet" in scientific reports and newspaper columns. At a distance of nearly 160 billion km (100 billion miles) from the Sun, and three times farther away than Pluto, 2003 UB313 is also the farthest object ever seen orbiting the Sun. It is also significant to note that the orbit of this "new planet" is even more non-circular than that of Pluto, and it takes about 560 years to make a single lap around the Sun.

The discovery of 2003 UB313 has raised extensive debate among astronomers as to what exactly defines a planet in the first place. Arguments have raged for years that because of its similarity to Kuiper Belt objects, Pluto should not actually be considered a planet. This viewpoint is probably counter to popular and cultural preferences, because this would decrease the number of planets to eight rather than raise the count to ten.

A committee of the International Astronomical Union, a special coordinating organization of professional astronomers, is now trying to decide how 2003 UB313 should be designated. If this object from the Kuiper Belt is formally declared to be a major new planet, another committee of the International Astronomical Union will have to decide what name it should be given. All of the other planets are named after Greek and Roman gods, and it's likely that this tradition would continue.

The root of the difficulty is that we currently do not have any scientific consensus about what makes an object a planet. If all objects larger than Pluto are to be designated planets, 2003 UB313 would indeed become the 10th planet of our solar system. However, this definition would also open up the possibility that there are a few more super-Plutos waiting to be discovered in the Kuiper Belt.

New Horizons

There is clearly a need for a better understanding of the objects at the edge of our solar system, and a new spacecraft mission offers exciting prospects for the future. In January 2006 NASA launched its New Horizons spacecraft to Pluto and its moon Charon. Pluto is the only planet that has so far never been explored by a mission launched from Earth. The New Horizon spacecraft mission is expected to reach Pluto by the middle of 2015, after gaining a gravity boost to its speed by swinging past Jupiter in early 2007.

The New Horizon spacecraft offers scientists a fantastic opportunity to study in close-up objects that are at the very outskirts of the solar system, including especially the surface properties and geology of Pluto. After a flyby past Pluto and its moon, the New Horizons spacecraft is scheduled for an extended mission travelling deeper into the Kuiper Belt. The promise here is for greater insights into "ice dwarfs", which can be contrasted with what we will have learned about Pluto. Great explorer missions such as New Horizons offer the real prospect of discovering icy bodies larger than Pluto in the Kuiper Belt. Undoubtedly, these findings will raise fresh controversy as to whether the planet count should rise. increase.

This is an artist's impression of the potential 10th planet in our solar system, 2003 UB313, looking back at the distant Sun. The inset shows a set of three discovery images from Palomar Observatory, California, taken over a period of three hours of 2003 UB313 (ringed). Larger than Pluto, the newly found body is seen moving very slowly against the "fixed" background of stars in these images. Much like Pluto, the cold and dark surface of 2003 UB313 is thought to be mostly covered in frozen methane.

A Torus of Rocks

The formation of the solar system was in some ways a rather messy process, with vast amounts of rock relics left over. These vagabonds have been striking and scarring the surfaces of planets and moons in violent impacts for billions of years. Included among these "travellers" are irregular-shaped rocky bodies known as asteroids, sometimes called "minor planets". Vast numbers of asteroids orbit the Sun at varying distances ranging from less than the radius of Earth's orbit to well beyond Saturn's path around the Sun. There is, however, a predominance of asteroids between the orbits of Mars and Jupiter, at about two and half times the distance between the Earth and the Sun. This torus, or doughnut-shaped structure, is known as the asteroid belt. It is a great ring of ancient rocky objects that encircles the four inner planets: Mercury, Venus, Mars, and Earth.

Tens of thousands of asteroids are known to reside in the asteroid belt, though it is estimated that there are actually millions. Most of the asteroids may be only the size of small rocks and only 16 of them are known to have diameters in excess of 240km (150 miles). The largest member of the asteroid belt is Ceres, which has a diameter of almost 950km (590 miles), about one quarter of the size of Earth's Moon. Only two other asteroids have diameters greater than 300km (186 miles) and approximately 200 are larger than 100km (62 miles). The majority of the asteroids in the belt are less than 1km (0.6 mile) across. Despite the high numbers of asteroids, the total mass of the asteroid belt is barely 12 percent of the Earth's Moon. If all of the asteroids were combined to form a single body, it would measure just 1500km (930 miles) across – less than half the diameter of the Moon. This means that the great swarms of belt asteroids are rather thinly spread and most of the belt is empty. It would be highly improbable for us to randomly pass close to an asteroid. Missions sent to the outer solar system, such as the Galileo and Cassini spacecraft, generally travel across the belt with minimal threat. The average distance between asteroids in the belt is an incredible 10 million km (6,200,000 miles). Spacecraft that actually travel close to asteroids have to be specially manoeuvred to pre-selected targets.

Since the 1990s, spacecraft have beamed to Earth close-up views of several asteroids. Images have even revealed that at least 30 asteroids, including Ida and Dionysus, have their own moons. The data have also uncovered a variety of compositions and characteristics. The vast majority of asteroids are thought to be composed primarily of stone (or silicates), while five percent are rich in iron and nickel. Although the irregularly shaped bodies tumble and spin, the rate at which they rotate is mostly unexpectedly low. It seems that repeated collisions over billions

The Galileo spacecraft's flight to Jupiter took it through the asteroid belt, where it made the first flybys of asteroids. This image shows the asteroid Ida from a distance of only 3500km (2175 miles). In addition to uncovering a heavily cratered body, the spacecraft also made the startling discovery that Ida has a 1.5km (0.9 mile) wide moon called Dactyl (seen far right). Close-up images of asteroids are advancing our understanding of the composition and origin of the asteroid belt.

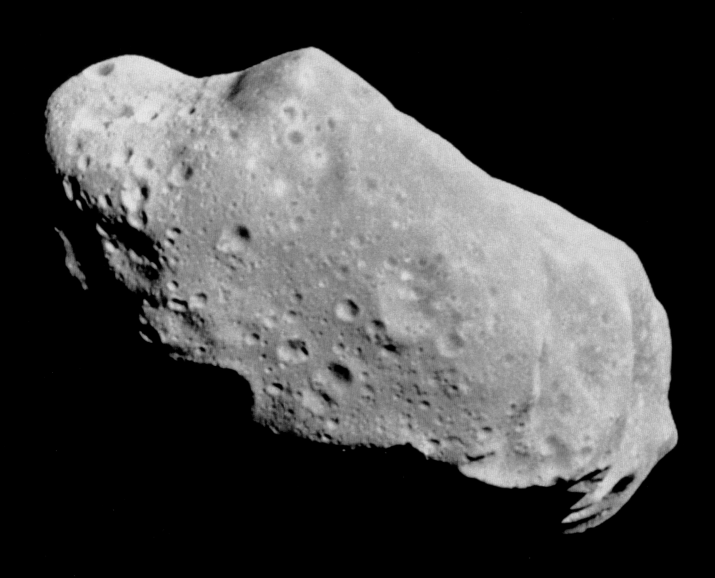

of years have reduced the asteroids to small piles of rubble, that are weakly held together by their own gravity. If they spun much faster, the asteroids would simply rip apart.

NASA is preparing a new space mission, Dawn, to study the two largest asteroids in the belt, named Ceres and Vesta. Due for launch in mid-2007, the Dawn mission will fly past Vesta first in 2011, and will arrive at Ceres four years later. The goal is to carry out a detailed study of the incredibly varied landscapes of these two asteroids, which are expected to include craters, canyons, and mountains. Scientists will use the data to advance

our understanding of the processes that occurred at the birth of the solar system. Another imaginative aspect of the Dawn mission is that the public were given the opportunity to submit their names to be recorded onto a microchip to be placed on board the spacecraft for the journey.

There are two scenarios for the possible origin of a well-organized concentration of asteroids orbiting in a belt around the Sun. Either they could be remnants of a planet that broke up billions of years ago, or perhaps they are rocks that failed to accumulate to form a planet. The current consensus among

Aan artist's idea of a ring of dusty debris surrounding an imaginary planetary system. Over hundreds of millions of years the planets orbiting a star sweep up most of the dust, leaving only a remnant ring of small, rocky bodies, similar to the Kuiper Belt that lies beyond the planet Neptune in our solar system.

astronomers is to favour the latter explanation. The total amount of material in the asteroid belt is too small to constitute a planet. Additionally, the asteroids do not all have the same chemical make-up, which suggests they cannot have originated from a single terrestrial source. Jupiter's strong gravitational field has played an important role in shaping the torus, or ring, of asteroid bodies. The giant gas planet acts to continually perturb the rocky members of the belt. This interference prevents the material in the belt from combining to form planets. Instead, the asteroids collide and fragment, with a substantial amount of the original matter possibly being ejected toward the planets or the Sun.

The asteroid belt is a primitive relic of the solar system. It has, however, been subjected to physical processes such as impacts, internal heating, and space weathering. These modifications mean that the Kuiper Belt objects much farther out in the solar system should be regarded as considerably more pristine, and therefore more representative of conditions in the earliest stages of the formation of the solar system, than those much nearer the Sun.

The discovery of planets beyond our solar system has become a great new pursuit of 21st-century science. Indirect methods have to date provided evidence of the existence of numerous planets, known as exoplanets, or extra-solar planets, around other stars. This artist's impression shows the view toward the first known planet to reside within a triple-star system. Located 149 light years away toward the constellation of Cygnus, the hot, Jupiter-like planet takes just over three days to orbit once around the most massive of the three stars. The scene depicts the planet at the upper left, viewed from the surface of a hypothetical moon.

Strange New Worlds

09-2005

One of the most exciting and vibrant areas of planetary study today concerns the hunt for distant planets orbiting stars other than the Sun, many light years away. Evidence exists for more than 150 extra-solar planets, or exoplanets, and the count continues to rise. In our current understanding of the process by which stars form, the creation of planets is a natural "by-product" forged from the left-over debris. The new technology that enables us to discover planet systems around other stars also causes us to question the uniqueness of our own solar system and the origin of planetary systems. New conceptual challenges are posed for example, by the observation of so-called "Hot Jupiters" orbiting closer to their parent stars than, for example, Mercury does to our Sun. The quest for the near future is to hunt out Earth-like planets around nearby stars. Impressive advances in observation techniques are being combined with technological advances to bring the realization of this goal much closer. The study of these new worlds at different stages in their histories provides clues about the evolution of our own planet, and its immediate neighbours. Even more remarkable is the prospect of directly detecting the weakly reflected starlight from Earth-like exoplanets, which will allow us to probe the chemical make-up of remote atmospheres, and perhaps even search out evidence of biological activity. It is exciting to contemplate that within a few decades we may know whether or not our appearance on Earth is indeed a unique quirk of nature and the Universe.

Our Milky Way is a spiral-shaped galaxy similar to NGC1309, captured here in stunning detail by the Hubble Space Telescope. The Sun is a fairly typical specimen of the billions of stars that are bound by gravity to form these magnificent structures. Given the vast numbers of stars, and the fact that the formation of planets is thought to be a natural by-product of the birth of stars, it is anticipated that our Galaxy may be teeming with planets.

Planet Foundries

Most astronomers agree that the formation of the solar system was a natural by-product of the formation of the Sun. The origin of our solar system dates back 4.5 billion years to the collapse under its own gravity of an enormous cloud of gas and dust. Most of the swirling material was drawn to the centre to form the Sun. The left-over debris was organized into a flattened disc orbiting the proto-Sun. The disc was loaded with dust-sized grains that collided with one another, occasionally sticking together to assemble clumps of matter. Over millions of years these clumps formed either the terrestrial, or rocky, planets, or the cores of giant gas planets. The hierarchical process of building planets from small to large bodies is today thought not only to be relevant to our solar system, but also to the formation of stars generally. Our Galaxy of two hundred billion stars held together in a magnificent spiral structure, may be teeming with extra-solar planets, also known as exoplanets, orbiting numerous Sun-like stars. This possibility of millions of planetary systems has driven astronomers to develop telescopes, detectors, and techniques to overcome the enormously difficult task of finding these planets.

One of the first challenges was to search for large discs of gas and dust around other very young stars. Known as proto-planetary disks, these ring-like structures extend billions of miles away from the stars and are believed to be embryonic planetary systems. These discs of debris not only show that star formation is well on its way, but they are also the foundries for the assembly of other planets. The discs indicate the presence of solid matter around a star in a flat structure that is very similar to the one out of which our own solar system condensed. Over the past few years, the sharp vision of the Hubble Space Telescope has clearly distinguished the new-born stars from their orbiting disks of dusty matter.

The highlights of recents observations include the debris disc imaged around a red-dwarf star called AU Microscopii. This star is only 12 million years old (our Sun is billions of years old). At a distance of only 32 light years from Earth, this star hosts the closest disk we have seen so far. Viewed edge-on it has a span of 64 billion km (40 billion miles). The powerful Hubble Space Telescope cameras have revealed kinks and distortions in the disc thought to be due to the gravitational influence exerted by planets forming, or already in existence. Another interesting case discovered in 2004 is the debris encircling the star named HD107146. This is a yellow-dwarf star very similar to, but younger than, our Sun. The disc imaged contains particles that are one hundred times smaller than household dust and in a greater abundance than they would have been in the disc originally surrounding our star.

Hubble Space Telescope observations have also revealed a spectacular dusty ring around a star called Formalhaut that is

A disk of dusty debris is seen in this detailed Hubble Space Telescope image of a star called HD 107146. The star is located 88 light years from Earth and is of a similar type to our Sun, although it is significantly younger. The disc represents an early stage in the formation of planets around this star and is essentially a vast reservoir of planet-building materials.

Overleaf
The main view shown is an artist's impression of the star AU Microscopii, with its dusty disc and a hypothetical planet. The inset upper-right is from the Hubble Space Telescope showing the disc around the star viewed nearly edge-on. Detailed analysis of this image has revealed that the disc is warped and shaped in a manner that is likely to be due to the gravitational influence of a planet.

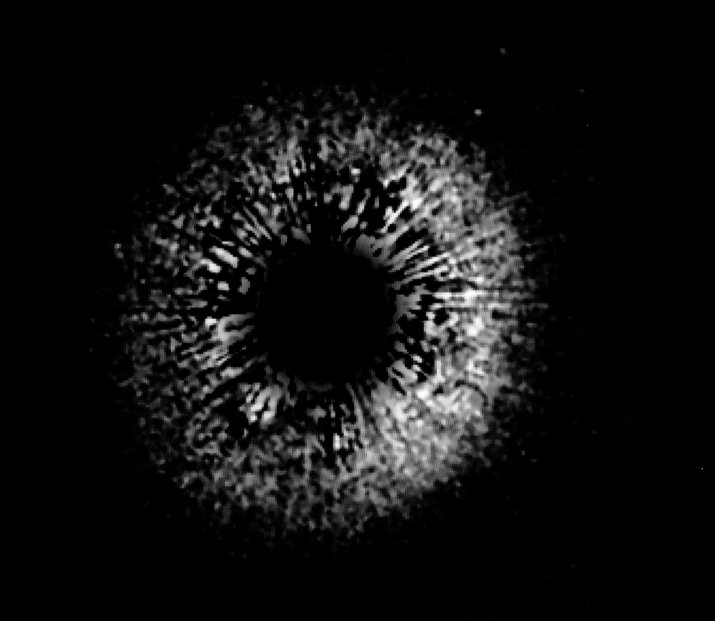

04-2004

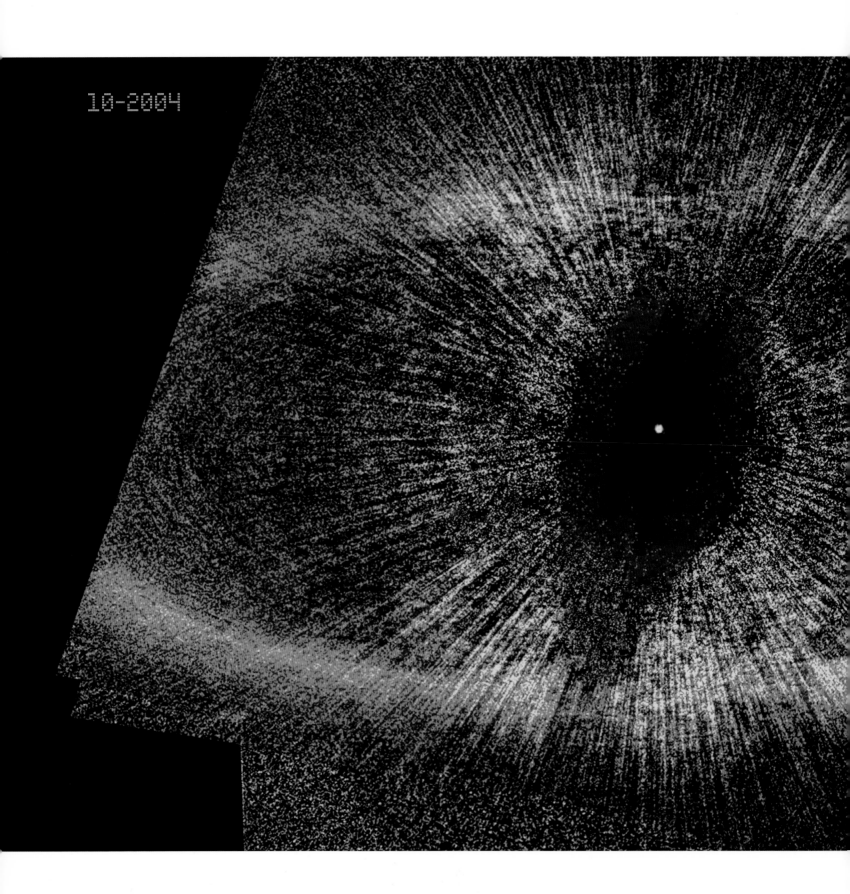

10-2004

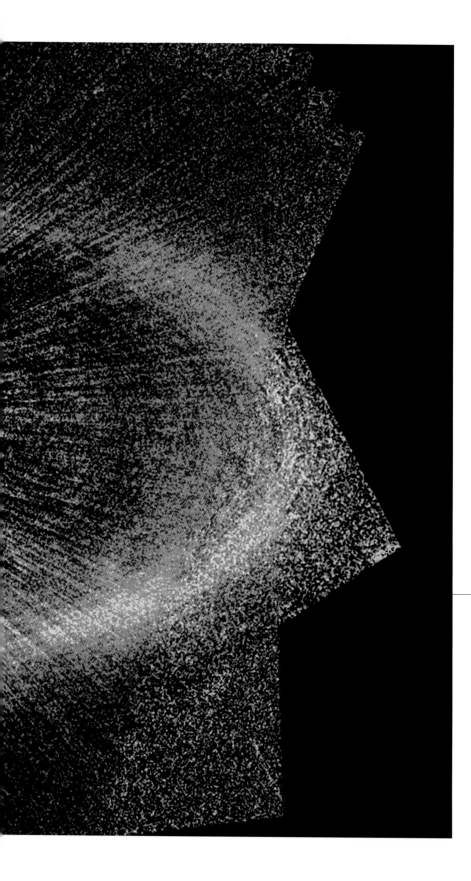

200 million years old. The diameter of the ring is four times greater than that of the Kuiper Belt in our solar system as the vast belt of icy bodies and dormant comets lying beyond the orbit of Pluto (see pages 154–7). The shape and structure of the Formalhaut disc suggest that an unseen planet has formed and is orbiting its parent star on an elliptical path. The gravitational pull created as the planet moves is reshaping the disc, while also helping to keep the ring material relatively narrow. An equally intriguing comparison to our solar system was revealed in 2005 by the infrared "eyes" of the Spitzer Space Telescope. Observations of a star named HD69830, 41 light years away from Earth, uncovered evidence of a massive asteroid belt where rocky Earth-like planets could be forged. Located toward the constellation of Puppis, HD69830 may help advance our understanding of the origin of terrestrial planets, because the giant asteroid belt is probably made up of the leftovers of the formation of such planets.

The fascinating observations of dusty debris discs around the youthful planets provide visual insight into the building blocks for their assembly. The exploitation of the infrared waveband region today is allowing us to peer through obscuring gas and dust to witness the very earliest stages in the formation of planets. The growing number of known proto-planetary discs suggests that it may indeed be easy to initiate the planet-building process, thus greatly improving the odds of our Galaxy being richly populated with planets. The next issue is whether we can find unambiguous evidence for the existence of exoplanets in star systems that have evolved beyond the stages of infancy.

Having blocked out the bright central starlight, this remarkable image from the Hubble Telescope reveals intricate details within a dusty ring of material surrounding the star Formalhaut. The distance between the star and the middle of the ring is about 15 times greater than the distance from the Earth to the Sun. The distortions apparent in the ring may be caused by the gravitational tug of a newly formed planet.

Overleaf
This pair of images shows significant stages in the lives of stars. The infrared image, left, from the Spitzer Space Telescope shows reservoirs of gas and dust that are the nurseries of newly formed stars. The tops of the 50 light-year pillars are lit by the prolific light from embryonic stars. In contrast, the image on the right, from the Hubble Telescope, shows the death of a massive star. It is a very high-resolution view of the Crab Nebula, the debris of hot gas after a supernova explosion. Planet-building and life-giving elements such as carbon and iron can be created only in the nuclear fusion engines of massive stars that die in this dramatic manner.

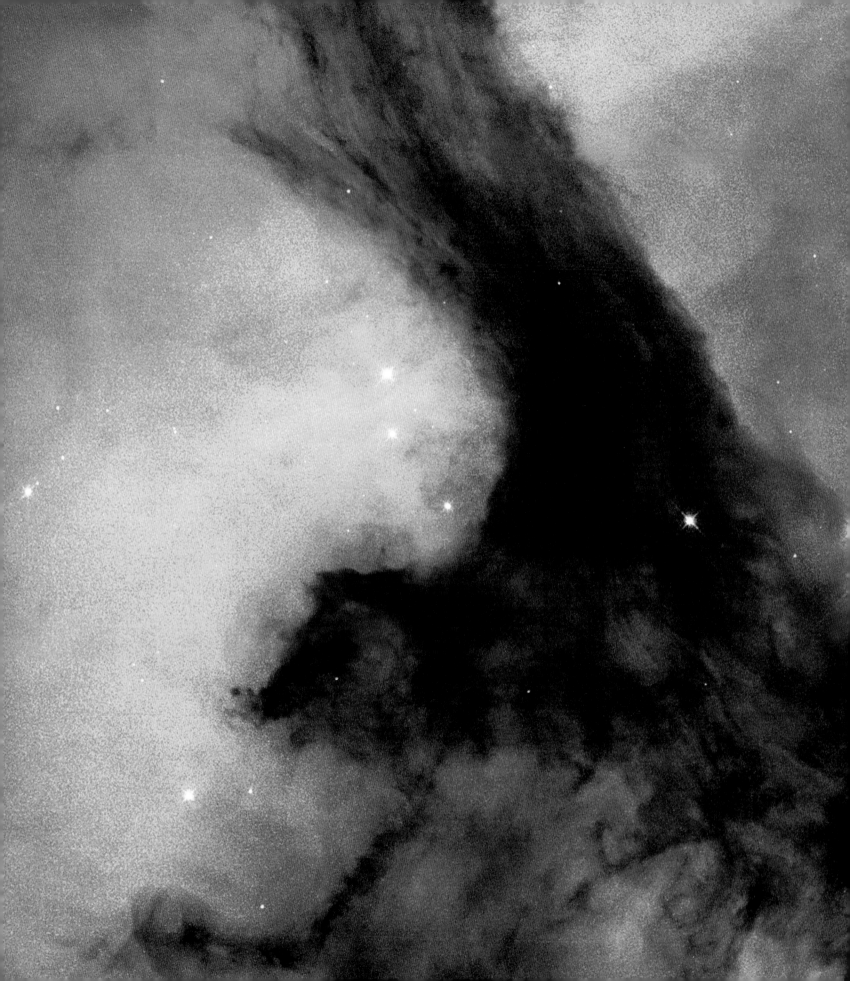

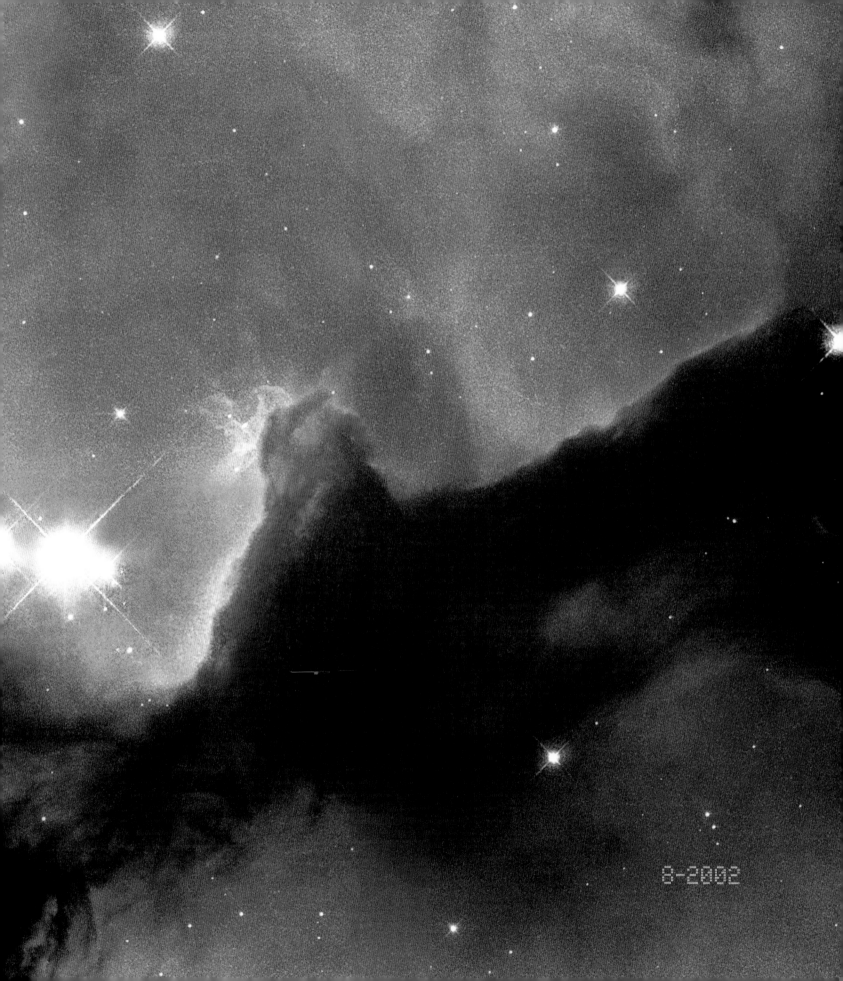

Hunting the Extra-solar Planets

Over the past decade more than 150 new planets have been discovered orbiting mostly dwarf stars that are comparable to the Sun in luminous power or size. These results of painstaking, dedicated, planet-hunting surveys are remarkable given that finding extra-solar planets, or exoplanets, is extremely hard. There are no pictures or images of these planets; instead their discovery is based on the indirect influence they have on the parent star. It is very difficult to see extra-solar planets because they are very small, lie close to their stars, and only faintly reflect the starlight. The Sun, for example, outshines its family of planets a billion times; quite simply, the feeble glimmer of a planet's light is lost in the outstanding glow of its star. An astronomer working at the location of the nearest star to our Sun, Proxima Centauri, would not be able to see any of the planets in our solar system, even with the sophisticated instruments and telescopes we have today. Because the other stars are much farther away than Proxima Centauri, astronomers have (for the moment) to look for indirect evidence for extra-solar planets.

The leading technique for planet hunting today relies on the fact that an orbiting planet exerts a gravitational tug on its parent star – as a planet revolves, it pulls the star first one way, then the other (creating a "wobble" effect). The bigger the planet, the more noticeable the wobble of the star. Extremely accurate measurements are made to analyse the wobbles in a star's motion to detect the presence of orbiting planets. These measurements provide information about the planet's mass, the period of its orbit, and its distance from the star. Our intrepid astronomer on Proxima Centauri would, for example, detect a tiny wobble in the Sun's motion with a 12-year pattern – the length of time it takes Jupiter to orbit the Sun once. The other planets in our solar system also cause the Sun to wobble, but by much smaller amounts. In fact, the movement caused by the Earth would be undetectable by our imaginary astronomer.

Through a regular monitoring programme, several research teams around the world have established how to measure the stellar wobbles accurately. The to-and-fro movement of a star is detected using very sensitive spectrographs connected to telescopes, which reveal a change, known as a Doppler shift, in the star's light. This is much the same as hearing sound waves, where an obvious example is the change in pitch of an ambulance siren as it approaches then recedes.

New planets are being detected at a rate of one per month, an avalanche of new discoveries being announced by groups of scientists working, for example, at the Lick Observatory, California and the University of Texas USA, University of Victoria, Canada, and Geneva Observatory in Switzerland.

An artist's impression depicting the growth of planetary systems through collisions (shown in the insets) between the small rocky bodies that reside in vast discs of debris left after the formation of a star. Planets are assembled over hundreds of millions of years, in a hierarchical process that begins with dust-sized particles, growing to rocky, or icy, bodies 1km (0.6 mile) or more wide, called planetesimals, and, ultimately, to terrestrial or gas planets.

Previous page
An astounding view obtained from the Hubble Space Telescope of the Trifid Nebula, which lies almost 9,000 light years away toward the constellation of Sagittarius. Dark bands of dust are seen against the background of glowing gas that forms this vast cloud in interstellar space. The Trifid Nebula is a prolific example of a site of star and planet formation in our Galaxy.

Another indirect, and in principle simpler, method of planet detection is known as the transit method. The idea is to scrutinize a large number of relatively nearby stars and look for alignments where an orbiting planet passes directly in front of the star as viewed from Earth. Somewhat like a miniature eclipse, the outcome is a "transit" in which the planet moves across the face of the star, blocking out a tiny fraction of the star's light. The resulting subtle decrease in the star's brightness is measured using telescopes. A similar effect occurs in our solar system when Mercury and Venus cross in front of the Sun, appearing to us as tiny black dots traversing a glaring circular disc. However, extra-solar planet transits are rare because they require a precise line up of the star, its planet, and our view from Earth. Nevertheless, the technique offers the reward of accurately assessing the size of the planet. If the planet's mass is also determined by measuring the star's wobble, the two measurements can be combined to provide an estimate of the planet's density.

Hot Jupiters

The vast majority of extra-solar planets discovered so far are gas giants. Their masses typically range from a shade smaller than Saturn's to more than 10 times larger than Jupiter's. The fact that these planets are so massive is not too surprising given that the detection techniques generally favour the discovery of large bodies over that of smaller ones. The newly found planets do, however, exhibit some unusual and unexpected characteristics. Unlike most of the planets in our solar system, which follow circular paths around the Sun, many of the extra-solar planets move in highly eccentric, oval orbits. What is even more peculiar is that, although the new planets may be much larger than Jupiter, they orbit extraordinarily close to their parent stars. In some cases they lap closer in than Mercury orbits our own Sun. For this reason these planets are known as "Hot Jupiters".

As a sub-set, Hot Jupiters have today become the most-studied extra-solar planets. An archetypical member of this group is a Sun-like star called 51 Pegasus, about four and a half

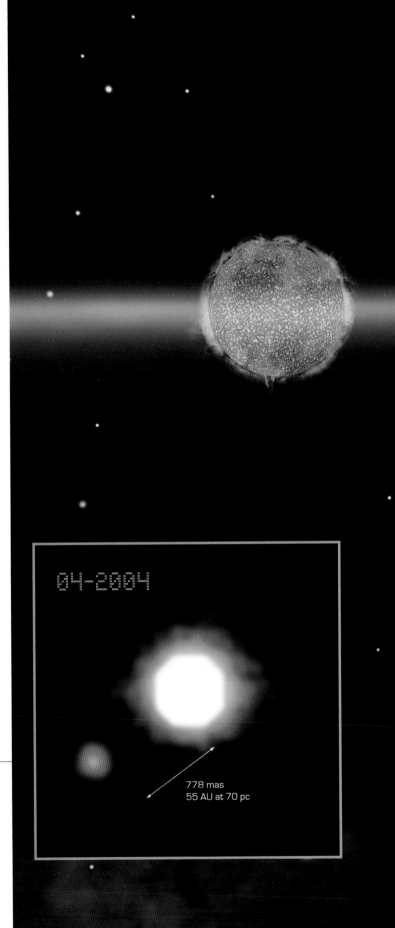

The feeble light from extra-solar planets is extremely difficult to detect. The inset bottom left is from the European Southern Observatory's 8m (26ft) class telescope in Chile. It is thought to reveal a large Jupiter-like planet (the faint reddish speck) orbiting a brown-dwarf star (centre right). The planet orbits nearly twice as far from its star as Neptune is from the Sun. The main view is an artist's rendering of the same extra-solar planet, which is almost five times larger than Jupiter.

778 mas
55 AU at 70 pc

light years from Earth. Measurements of this star's wobble indicate that it is orbited by a planet roughly half the mass of Jupiter, but at barely 0.05 times the distance between the Earth and the Sun. Mercury lies between six to eight times further away from the Sun than it does. Other stunning examples of Hot Jupiters include a planet orbiting so close to the star Tau Bootis, that its surface temperature is more than 1200°C (2192°F). A star called HD195019 hosts a planet more than 1000 times larger than Earth but almost eight times closer in. Some of the giant gas planets are so close to their stars that they whiz around their full orbits in less than three Earth days.

An even hotter pair of planets was seen by astronomers in 2004 using the powerful 8m (26ft) telescopes of the European Southern Observatory's facility in Paranal, Chile. The Jupiter-sized bodies orbit their stars in less than two days and are more than 15 times closer to their stars than Mercury is to the Sun. These very hot planets probably have temperatures in excess of 1800°C (3270°F), making them, and other giant extra-solar planets, extremely unlikely to be habitable. Their very hot atmospheres will be far more turbulent than any of the stormy conditions on Jupiter. It is intriguing to speculate that potentially large terrestrial moons may orbit these exoplanets, offering more hospitable conditions for biological organisms.

The great puzzle is how such massive planets end up so remarkably close to their stars. Their discovery defies the conventional theory for the formation of planets in our solar system, which states that giant gas planets such as Jupiter and Saturn form in the icy outskirts of the original collapsing disk of gas and dust. On the basis of our current knowledge, the giant extra-solar planets should be farther away from their stars and in nearly circular orbits. The new discoveries have forced a re-think; the consensus today is that the Hot Jupiters probably did form much farther out, but they then migrated inward toward their stars. The movement is probably due to a gravitational interplay between the newly formed planets and the dusty disc out of which they coalesced. This is similar to the way that Saturn's moons distort the structure of the planet's rings, setting up complex spiralling waves of material. The fact that Jupiter's atmosphere contains higher proportions of heavy gases, such as krypton and xenon, than expected is being interpreted as evidence that our giant gas planets may have moved to their present positions from even more remote locations in the early evolution of the solar system.

The movement of these exoplanet giants, however, has consequences for any small terrestrial planets. The closer a massive, Jupiter-sized planet is to its host star, the smaller the chance that Earth-like planets will form in that star system. This is similar to the way that Jupiter's gravitational influence may have hindered the material in the asteroid belt from collecting to form a single body. Another possible implication is that the inward migration of big giants could bulldoze small rocky planets, either in toward their star or completely out of the planetary system. It is clear that Earth-like planets are more likely to be found in multiple-planet systems where the giant planets are perhaps not so close to the star. A good example is the yellow Sun-like star 47 Ursae Majoris, which is located 51 light years

An illustration of the proposed Kepler spacecraft in orbit. Due for launch in June 2008, this satellite will uniquely conduct a census of Earth-sized planets around other stars. It will monitor almost 100,000 stars in a hunt for terrestrial planets that are orbiting their parent stars in the critical habitable zone, where the temperature is favourable for liquid water to exist on a planet's surface.

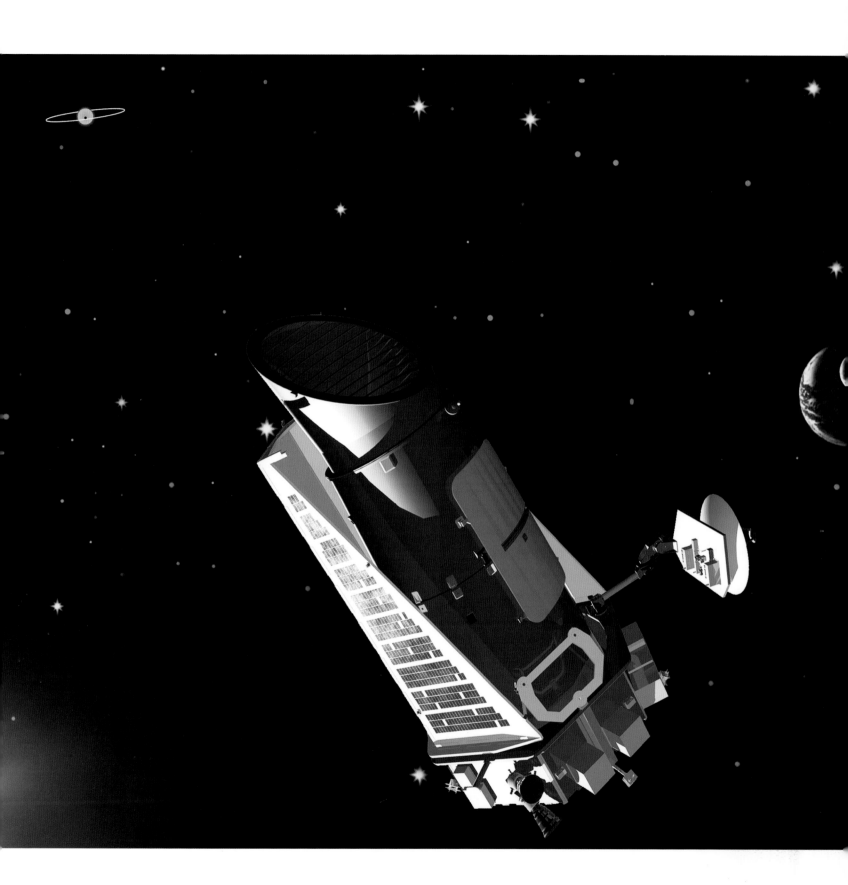

away towards the constellation of the Big Bear (or Big Dipper). Two planets with nearly circular orbits have been detected around this star, of comparable sizes to Jupiter and Saturn. Interestingly, if these planets were placed in our solar system, they would be located beyond the orbit of Mars. This leaves a large inner gap for possible small rocky planets. The discovery of multiple-planet systems in general has increased speculation that smaller, Earth-like planets may one day be uncovered.

Homing in on Earth-like planets

In recent years, as more sensitive detection equipment has been developed, smaller and smaller extra-solar planets have been discovered, typically down to the size of Saturn and Neptune. The ultimate holy grail of discovering an Earth-like planet orbiting a Sun-like star demands more sophisticated search techniques, and advances in space-based observatories.

In January 2006, astronomers using a relatively new planet-searching method announced the discovery of the smallest planet so far found orbiting beyond our own solar system. The object was discovered using a method known as gravitational microlensing, based on a concept originally studied by Albert Einstein. Looking toward a densely packed background of stars, very occasionally we see that one of these stars has its light amplified by a microlensing effect, whereby its light is bent around a precisely aligned foreground star. The gravity of the foreground star acts as a "lens" to create a powerful natural telescope. The background star will appear to brighten and then fade over a period of several days, or weeks. If the foreground star has a planet, the "lens" acts differently in that a second smaller amplification of the background star occurs over a shorter time-scale. Using a network of robotically controlled telescopes, an international collaboration of astronomers used the gravitational lensing technique to uncover a planet barely five and half times the size of our planet Earth. This small rocky exoplanet is orbiting a star 20,000 light years away in the constellation of Sagittarius, towards the centre of our Milky Way Galaxy. The planet is estimated to orbit about three times farther away from its star than the Earth does. It is likely to be a frigid, ice-covered body with a surface temperature of -220°C (-365°F).

The newly discovered rocky planet is probably too cold to host biological organisms, but it does provide evidence to support the view that low-mass extra-solar planets may be considerably more common than Hot Jupiters. Our success in finding smaller planets boils down to developing more sensitive methods than the "wobble" technique used to discover the majority of planets so far. An outstanding prospect over the next few years is that we may be able to actually "see" the light directly from small terrestrial planets orbiting tens of light years away. NASA and the European Space Agency are actively pursuing projects that will exploit networks of telescopes in space to image exoplanets. The big challenge is to find terrestrial planets that are up to 600 times smaller than Jupiter.

One of the most exciting space missions we can look forward to in the near future is NASA's Discovery-class mission called Kepler, after the 17th-century German astronomer Johannes Kepler, who was the first person to correctly explain the motion of the planets in the solar system. The spacecraft will use a purpose-designed 0.95m (3ft) wide telescope to monitor the brightness of more than 100,000 stars in our Galaxy for the tell-tale brief dips in their light that occur when planets pass in front of their host stars. These planet transits will produce a pattern of minuscule brightness changes that should repeat on a precise clock related to the orbit of the planet. Scientists anticipate that the Kepler mission will detect about 50 planets around the size of the Earth and almost 600 planets of up to twice that size. Depending on funding constraints, this ambitious project is due for launch in June 2008 and will explore an incredible diversity of planet systems.

Although the Kepler mission stands to advance our knowledge considerably, it will, however, still not directly image the planets themselves. The presence of the planets will be inferred from the temporary dimming of light resulting from transits. The possibility of detecting Earth-like extra-solar planets through imaging demands even greater telescope power and remarkably sensitive imaging technologies. The Terrestrial Planet Finder is a new space mission being designed to meet this challenge. It is envisioned to use a suite of 3.5m (11ft 6in) telescopes in space, which will work together and operate in the infrared region of light. By coordinating these powerful space telescopes, the Terrestrial Planet Finder will be able to detect the feebly reflected starlight and heat energy from terrestrial planets as they orbit nearby Sun-like stars. With a planned launch date of 2014, this revolutionary mission will exploit the results from the Kepler spacecraft to select the most appropriate targets for observation. Although the images captured by the Terrestrial Planet Finder will be of rather low resolution, they will be of sufficient quality to allow the observatory to analyze the faint light received for its chemical properties. Using a spectrogram, scientists will be then able to test the content of the planets atmospheres and determine whether a planet is habitable, or even whether it has life. There is a thus need to understand biosignatures not just in our world, but also on other more hostile locales in the solar system. The instruments on board the Terrestrial Planet Finder will be used to probe the atmosphere of Earth-like planets for traces of gases such as carbon dioxide, water vapour, and ozone. On Earth, for example, the only mechanism for continually maintaining atmospheric ozone and, by implication, oxygen, is life.

Beyond missions such as the Terrestrial Planet Finder, astronomers anticipate even more powerful telescopes in space or on the Moon. Spread over an effective operating span of hundreds of miles, these telescopes would work in harmony to obtain clearer images of other Earth-like planets. The greater sensitivity and resolution of grander future space missions will permit the most probing search for biosignatures and extend the hunt for new worlds to many thousands of stars. Within the next two to three decades we stand to uncover whether our Earth is truly rare, or rather common, and with this realization we can achieve scientific advances that will have enormous philosophical implications. It is possible that our children will, within their lifetimes, be captivated by images of oceans, continents, and mountains as the new wonders of the exoplanets.

Overleaf
Peering into just a tiny region of space, the Hubble Space Telescope revealed this remarkable collection of hundreds of nearby and distant galaxies. This grand scene is a reminder of the mediocre place of our own solar system, which is peripherally located in the outer region of our Galaxy. We inhabit a 14-billion-year-old Universe that contains hundreds or even billions of galaxies. The Universe is also constantly changing as planets and stars are born and die.

Glossary

acceleration – manner in which speed and direction of motion changes with time.

amino acids – molecules that form the building blocks of proteins.

archaea – one of the current highest levels for the classification of life.

astrobiology – the scientific study of life beyond Earth.

asteroid – a small rocky body in space. Most of the several thousand asteroids in the Solar System reside in the "belt" that lies between the orbits of Mars and Jupiter.

atom – the building blocks of matter. Each element (such as hydrogen or helium) is characterized by a unique type of atom.

aurora – a light display that results when charged particles from the Solar wind enter the Earth's atmosphere around the North or South Poles. Sometimes called the northern or southern lights on Earth.

axis – a straight line about which an object rotates or orbits.

basalt – dark, dense rock commonly produced by undersea volcanoes.

black smokers – outpourings around sea-floor vents that support life on Earth.

climate – a description of the long-term average of weather.

comet – a small body made mostly of dust and ice, travelling in an elliptical orbit about the Sun. As a comet nears the Sun, some of its material boils off to form a long tail.

compound – a substance that can be further broken down into elements.

constellation – a grouping of stars identified on the sky. Many constellations are named after characters or beasts from ancient mythology.

core – referring to the central part of an object, such as planet, star, or galaxy.

corona (of Sun) – the tenuous and extremely hot outer atmosphere of the Sun.

coronal mass ejection – a transient event in which billions of tons of gas are suddenly blasted away from Sun into space.

crater – a bowl-shaped depression on the surface of a planet or moon, resulting from the impact of a smaller body.

crust – the outer surface layer of a terrestrial planet.

density – the amount of mass contained in a certain volume.

disc – a flattened, rotating structure of gas and possibly dust.

Doppler Shift – the change in colour of light when the source of the light and the instrument of observation are moving with respect to one another. An analogous effect in sound is the way the pitch of a whistle changes as a train passes by.

dust – in astronomy, microscopic bits of solid matter found in space.

eclipse – an event during which one object passes in front of another.

electromagnetic spectrum – the complete range of light as characterized by wavelength, frequency, or energy. Examples include X-ray light, ultraviolet light, visible light, infrared light, and radio waves.

electron – a negatively charged sub-atomic particle that normally moves about the nucleus of an atom.

element – a substance that cannot be reduced by chemical processes into a simpler substance.

erosion – wearing down of geological features due to the action of water, wind, ice, and so on.

exobiology – the study of life as it may occur elsewhere than on Earth.

exoplanet – see extra-solar planet.

extra-solar planet – a planet orbiting around a star other than the Sun.

evaporation – the change in phase from liquid to gas.

fossil – an ancient relic of a living organism that died long ago.

flare – an explosive event occurring in or near an active region on the Sun.

force – action on an object that causes its momentum to change.

fusion (nuclear) – the process by which light atomic nuclei are combined into heavier ones, with a release of energy.

galaxy – a collection of millions and billions of gravitationally bound stars, plus gas and dust.

Galaxy – the galaxy to which the Sun belongs; same as the Milky Way Galaxy.

gravity – the mutual attraction between objects with mass. The greater the mass of a body, the stronger its gravitational pull.

helio – a prefix referring to the Sun.

hypothesis – an idea or concept that is put forward but remains to be tested with experiment or observation.

impact crater – see crater.

infrared radiation – the part of the electromagnetic spectrum just beyond the red-end of the range that can be seen by the eye.

ion – an atom with a negative or positive electrical charge.

jet – a stream of high-speed gas and particles ejected in a confined or beam-like manner.

jovian planet (giant planet) – a large planet resembling Jupiter that is mostly composed of gas (as opposed to rock and metal substances).

lander – a spacecraft or probe that lands on the surface of another body, such as a planet or moon.

lava – molten rock flowing on the surface of a planet.

light year – the distance travelled by light in a vacuum in one year. One light year is equal to 9460 billion km (5880 billion miles).

magma – molten rock beneath the surface of a planet.

magnetic field – a region of space associated with a magnetized object within which magnetic forces can be detected.

mantle (of Earth) – a layer of the Earth's interior that lies above the core and just below the crust.

mass – a measure of the amount of matter contained within a body.

metabolism – collective chemical reactions that occur in living organisms.

metal – in astronomy, an element other than hydrogen or helium. In the context of planets, it has the more conventional meaning, referring to substances that are good conductors of electricity, such as iron, tin, and so on.

meteorite – any remains from a meteor that survives passage through the atmosphere and strikes the ground.

microbe (microbial) – an organism that is so small that it can be seen only with a microscope.

micro-organism – see microbe.

Milky Way – the band of light that encircles the night sky, due to the numerous stars and nebulae lying close to the disc of our Galaxy.

minerals – solid compounds that form rocks.

model – a theoretical construct used to explain an observation or experiment.

molecule – a particle that results from the combination of two or more atoms tightly bound together.

momentum – measure of the state of motion of a body, defined as the product of its mass and velocity.

nebula – a cloud of gas and dust in space.

nuclear – referring to the nucleus of an atom.

optical – relating in astronomy to the visible-light band of the electromagnetic spectrum.

orbit – the path of one astronomical body about another body or point.

orbiter – a spacecraft that goes into orbit around another solar system body.

outgassing – the release of gases into a planet's atmosphere due to volcanic activity.

ozone – a layer of gas in the atmosphere of an Earth-like planet at a height of tens of kilometres above the surface, where incoming ultraviolet radiation from the Sun (star) is absorbed.

planet – a large body (rocky or gaseous) that orbits a star.

planetesimal – a small body of dust and ice from which planets are formed.

plasma – a gas that is fully or partially ionized.

pressure – refers to how much force is spread over a given area.

proton – a sub-atomic particle with a charge opposite to an electron but having a much greater mass.

proto-planet – a planet that is in the process of forming.

radar – a technique of bouncing radio waves off an object and then detecting the radiation that the object reflects back to the radio transmitter.

radiation – usually refers to electromagnetic radiation, such as visible light, infrared, ultraviolet, and so on.

reaction – a process that involves changes in the structure and energy content of atoms and molecules.

ringlets – narrow bands of particles that make up Saturn's system of rings.

sediment – deposits and cementation of fine grains of material usually resulting from erosion in lakes and oceans.

shepherd moon – a moon whose gravity affects the movement of particles in a planetary ring.

solar system – the system of bodies that are gravitationally bound to the Sun, such as the planets, moons, comets, and asteroids.

spectrograph – an instrument attached to a telescope to create an image of a spectrum.

spectrogram (spectrum) – the array of colours or wavelengths apparent when light is dispersed, for example by a prism.

speed – a measure of motion in terms of distance travelled over time.

spiral galaxy – a type of galaxy in which most of the gas and stars are in a flattened disc that displays spiral arm structures. Our Milky Way Galaxy is a spiral galaxy.

star – a massive sphere of gas that shines by generating its own power.

Sun – the star about which the Earth is orbiting.

temperature – a measure of how hot or cold an object is; a measure of the average random speeds of microscopic particles in a substance. Celsius, Farenheit – temperature measured on a scale in which water freezes at $0^{o}C$ ($32^{o}F$) boils at $100^{o}C$ ($212^{o}F$).

terrestrial planet – a planet that is predominantly composed of rocky and metal substances. Earth, Venus, and Mars are terrestrial planets.

theory – a set of laws and hypotheses has have been used to explain observed phenomena.

thermonuclear energy – energy that results from encounters between particles that are given high velocities (through heating).

tidal force – the differences in the force of gravity across a body that is being attracted by another, larger body. The result may be the deformation or destruction of the smaller body.

ultraviolet radiation (light) – electromagnetic radiation of the region just outside the visible range, corresponding to wavelengths slightly shorter than blue light.

Universe – the total of all space, time, matter, and energy.

vent – a hole or opening in the crust (perhaps sea floor) of a planet.

visible light – see optical.

volatile – an element with low melting and boiling temperatures.

volcano – an eruption of hot lava from below a planet or moon's crust to the surface.

volume – a measure of the total three-dimensional space occupied by a body.

waveband – a general term for a broad region of the electromagnetic spectrum, such as ultraviolet, X-ray, and optical.

wavelength – the distance between two successive peaks or troughs of a wave.

weight – the total force on some mass due to gravitational attraction.

X-rays – high-energy electromagnetic radiation, with photons of wavelengths intermediate between those of ultraviolet radiation and gamma rays.

zonal winds – alternating eastward and westward winds in the atmospheres of giant gas planets.

index

acknowledgments

9 SOHO/NASA, 10-11 NASA/JPL-Caltech/E Churchwell (University of Wisconsin), 12 NASA/Stardust, 14-15 NASA/JPL/Texas A&M University/Cornell University, 16-17 NOAO/AURA/NSF, 18-19 ESA, 20 NASA, 23 Science Photo Library/NASA/ESA/STScI, 24 JPL/NASA, 25 Science Photo Library/NASA/ESA/STScI, 27 NASA/JPL, 28 Science Photo Library/NASA, 29 NASA/JPL/Cornell University, 30-31 NASA/JPL/Space Science Institute, 32 main ESA/DLR/FU Berlin (G Neukum), 32 insets NASA/JPL/MSSS, 35 NASA/TRACE, 36 SOHO/EIT and SOHO/LASCO (ESA & NASA), 38-39 Image Science & Analysis Laboratory, NASA Johnson Space Center, 40-41 NASA/ESA/J Clarke (Boston University) and Z Levay (STScI), 42-43 Science Photo Library/NASA/ESA/STScI, 44-45 NASA/NSSDC, 46 NASA/JPL/NIMA, 49 NASA/GFSC/METI/ERSDAC/JAROS and US/Japan ASTER science team, 50 NASA/JPL/NGA, 53 E De Jong et al (JPL)/MIPL/Magellan Team/NASA, 54-55 NASA/JPL-Caltech, 57 NASA/JPL/MSSS, 58-59 ESA/DLR/FU Berlin (G Neukum), 61 NASA/JPL, 62, 64 above left & right, below left NASA/JPL/University of Arizona, 64 below right NASA, 66 Science Photo Library/NASA/ESA/STScI, 67, 68 left & right, 69 NASA/JPL/Space Science Institute, 70-71 NASA/JPL/US Geological Survey, 72-73 ESA/DLR/FU Berlin (G Neukum), 74 NASA/JPL/Space Science Institute, 76-77 NASA/JPL/Cornell University, 78-79 NASA/JPL/Space Science Institute, 81 NASA/JPL/MSSS, 82-83 NASA/JPL/Cornell University, 84-85 NASA/JPL, 87 Science Photo Library/ESA, 88-89 Science Photo Library/NASA, 90 left NASA/GSFC/LaRC/JPL, MISR Team, 90 right NASA/JPL, 91 From *Terrestrial Impact Craters*, second edition, compiled by Christian Koeberl and Virgil L Sharpton, Lunar and Planetary Institute (LPI), contact webmaster@lpi.usra.edu, 92 NASA/JPL/Space Science Institute, 93 Science Photo Library/NASA, 94-95 NASA/JPL/US Geological Survey, 96-97 NASA/JPL/Caltech, 99 NASA/JPL-Caltech/UMD, 100 Hubble Space Telescope Comet Team, 101 Dr H A Weaver and T E Smith (STScI/NASA), 102 Hubble Space Telescope Comet Team/NASA, 103 SOHO/SWAN & SOHO/LASCO (ESA & NASA), 106-107, 108-9 NASA/JPL, 110 NASA/NOAO/NSF/T Rector (University of Alaska, Anchorage), Z Levay and L Frattare (STScI), 113 NASA, 114 NPS, photo by A Mebane, 115 OAR/ National Undersea Research Program (NURP)/NOAA, 116-117 Science Photo Library/Space Imaging, 118-119 NASA/JPL/MSSS, 120 NASA/JPL/Goddard Space Flight Center, 121 NASA/JPL, 122-123, 125, 126-7 ESA/DLR/FU Berlin (G Neukum), 129, 130 left NASA/JPL/Space Science Institute, 130 right NASA/JPL/University of Arizona/University of Colorado, 131 NASA/JPL/University of Arizona, 132 left taken from Cassini's Three Views of Titan: NASA/JPL/Space Science Institute, 132 right NASA/JPL/Space Science Institute, 135 NRAO/AUI/NSF, 136-137 Science Photo Library/NASA, 138 NASA/JPL/Space Science Institute, 141 NASA and the Hubble Heritage Team (STScI/AURA), acknowledgment: R G French (Wellesley College), J Cuzzi (NASA/Ames), L Dones (SwRI), and J Lissauer (NASA/Ames), 142-143 NASA/JPL/Space Science Institute, 144 left NASA/JPL/University of Colorado, 144 right, 146 below right NASA/JPL, 146 above & below left, above left & right, NASA/JPL/Space Science Institute, 148-149 Science Photo Library/NASA/ESA/STScI, 150 above & below, 151, NASA/JPL, 152-153 NASA/JPL/Caltech, 155 NASA/JPL, 156 main NASA/JPL/Caltech, 156 inset Samuel Oschin Telescope, Palomar Observatory: Mike Brown (Caltech), Chad Trujillo (Gemini Observatory), and David Rabinowitz (Yale University), 159 Science Photo Library/NASA, 160-161 NASA/JPL-Caltech/T Pyle (SSC), 162-163 Caltech, 164 NASA/ESA/the Hubble Heritage Team, (STScI/AURA) and A Riess (STScI), 170-171 NASA/ESA/P Kalas and J Graham (University of California, Berkeley) and M Clampin (NASA/GSFC), 177 NASA/JPL-Caltech/T Pyle (SSC), 167 Science Photo Library/NASA/ESA/STScI, 168-169 main NASA/ESA/G Bacon (STScI), 169 inset NASA/ESA/J E Krist (STScI/JPL), D R Ardila (JHU), D A Golimowski (JHU), M Clampin (NASA/Goddard), H C Ford (JHU); G D Illingworth (UCO-Lick); G F Hartig (STScI) and the ACS Science Team, 172 NASA/JPL-Caltech/L Allen (Harvard-Smithsonian CfA), 173 NASA/ESA/J Hester and A Loll (Arizona State University), 174-175 NASA/ ESA/the Hubble Heritage Team (AURA/STScI), 177 SA/JPL-Caltech/T Pyle (SSC), 178-179 & inset ESO, 181 NASA, 184-185 NASA/ESA/the Hubble Heritage Team (STScI/AURA).